quake

Hawke's Bay 1931

Matthew Wright

Also by the author:
Hawke's Bay: Lifestyle Country
Wonderful Wairarapa
Wellington/Kapiti Coast
Hawke's Bay – the history of a province
Havelock North – the history of a village
Napier – city of style
Kiwi Air Power – the history of the RNZAF
Working Together – the history of Carter Oji Kokusaku Pan Pacific Ltd 1971Ð1993
New Zealand's Engineering Heritage
Battle for Crete – New Zealand's near-run affair, 1941
Quake – Hawke's Bay 1931
Town and Country – the history of Hastings and district
Blue Water Kiwis – New Zealand's naval story
Desert Duel – New Zealand's North African war
Wings Over New Zealand – a social history of New Zealand aviation
Italian Odyssey – New Zealanders in the battle for Italy 1943-45
Rails Across New Zealand – a social history of rail travel
Pacific War – New Zealand and Japan 1941-1945
The Reed Illustrated History of New Zealand
Western Front: the New Zealand Division in the First World War
Freyberg's War: the man, the legend, and reality
Cars Around New Zealand

Front cover image courtesy Rene Hare.
Back cover image courtesy Storkey Collection.

Established in 1907, Reed is New Zealand's largest
book publisher, with over 600 titles in print.
www.reed.co.nz

Published by Reed Books, a division of Reed Publishing (NZ) Ltd, 39 Rawene Road, Birkenhead, Auckland 10. Associated companies, branches and representatives throughout the world.

A catalogue record for this book is available from the National Library of New Zealand.

ISBN-10: 0 7900 0776 2
ISBN-13: 978 0 7900 0776 2
First published 2001
Reprinted with changes 2006

Cover designed by Cathy Bignell

Printed in New Zealand

Contents

Introduction to the 2006 edition

The Hawke's Bay earthquake remains New Zealand's quintessential twentieth century disaster, an arbiter of town character in both 'Art Deco' Napier and 'Spanish Mission' Hastings. On a per-capita basis it was New Zealand's most lethal single calamity, and the absolute number killed and injured on that single day in February 1931 runs second only to losses during the New Zealand attack on Passchendaele, 14 years earlier. Some Hawke's Bay people lived through both; and as we will see in these pages, the human response to the quake was shaped by memory of the First World War.

This edition of *Quake – Hawke's Bay* 1931 includes over 40 new pictures, including a selection of quake-era images published for the first time. Some are from the family collection of the late Reverend Douglas Storkey. Others are from a private set taken a few days after the quake. To this are added illustrations from the collection of the Alexander Turnbull Library not used in the original edition; and a series of images from the contemporary Napier townscape.

This book incorporates documentary material collected over a number of years. I am indebted to the late Ken Hawker for his generous provision of photographs from the *Daily Telegraph* collection; and to Paul Taggart and *Hawke's Bay Today* for kind permission to use both these pictures and extracts from Before and After. I am grateful to the late Douglas Storkey for permission to use photographs from the W.E. Storkey souvenir collection, and for his kind provision of pictures from his family collection. I also thank R.O. Hare, and Ross and Jill Johnson for their generous loan of images from the quake. The Hawke's Bay Cultural Trust, Havelock North Public Library, Hastings Public Library and the staff of the manuscripts and photographic sections of the Alexander Turnbull Library provided invaluable assistance. My particular thanks go to my wife Judith, for her patience with my anti-social weekend writing habits.

Matthew Wright
July 2005

Introduction

The Hawke's Bay earthquake of 3 February 1931 came like a 'bolt from the blue',[1] a sudden catastrophe unpredicted by the science of the day. It turned the province into a virtual war zone, and remains New Zealand's most destructive single disaster. Its impact can be seen in the letters, diaries, memoirs and photographs of rescuers working desperately to pluck victims from wreckage before all was consumed by fire, the doctors and nurses who tended the injured in makeshift hospitals, and the refugees walking down broken roads with their meagre belongings balanced precariously on bicycles. It seared itself indelibly into the memories of the 70,000-odd men, women and children living in Hawke's Bay that fateful summer.[2]

At one early stage authorities feared that up to 300 people had died.[3] The toll was finally settled officially at 256 — although, despite an outstanding effort, two people were never accounted for.[4] Over 400 were hospitalised with serious injuries. At least 2500 were cut, bruised, scratched and shocked, a toll never fully evaluated because few bothered to report trivial injuries to the authorities. The earthquake was an indiscriminate killer; its victims ranged from six-week-old David Tripney to 92-year-old Gilbert Brown.[5] The total number of injuries and fatalities on that sunny February day exceeded those of

Photographer A.B. Hurst rushed from his studio to snap pictures as the vibrations subsided. His pictures are the only record of Napier after the quake but before the fire that reduced much of the central business district to total ruin. This image is from a selection purchased by Napier bookseller W.E. Storkey for publication in souvenir sets, which were widely sold to both local public and tourists from as far afield as the United States. Within a few hours, the Emerson Street wreckage seen here had been obliterated by fire. (STORKEY COLLECTION)

any other single national disaster by a wide margin.[6] The deeper implication was even more grievous; such heavy destruction during the second summer of the depression threatened complete ruin.

The event is often called the 'Napier' or 'Hawke's Bay' earthquake, but it actually affected the whole of depression-era New Zealand. The shock brought down buildings between Gisborne and Waipawa. It toppled chimneys from Taupo to Wellington. The whole country rallied to assist the beleaguered residents of Hawke's Bay. News of the calamity even depressed stocks on the London exchange.[7] The building standards established afterwards set the pattern for the rest of the century, and the economic impact became a model for studies of the effects of the feared Wellington 'big one' into the twenty-first century.[8]

Ongoing studies during the last years of the twentieth century have altered every previously accepted detail of the seismic event, ranging from the magnitude — revised down in 1981 to a level of 7.8 — to the location of the 'rupture zone', the huge slab of landscape that actually moved.[9] At the human level, diaries, photographs, eyewitness accounts and memories paint a picture of horror that spans the decades. New material demands new analysis. As Eric Hobsbawm has noted, historians aim 'not to discover the past, but to *explain* it ... what we want to know is *why*, as well as *what*'.[10] The story of the Hawke's Bay earthquake is no exception.

CHAPTER 1

'A bolt from the blue'[1]

Very heavy and severe earthquake shocks taking place in Napier. 2/t clock stopped. Vessel shaking and vibrating violently from stem to stern.

RADIO LOG, SS *TARANAKI*, 2247 HOURS GMT, 2 FEBRUARY 1931 (10.47 A.M., 3 FEBRUARY LOCAL TIME)[2]

Oh! It was terrible — the shrieks of the wounded, the moaning of the dying, and the terror in the eyes ...

MISS A. SAMPSON, HASTINGS[3]

In Hawke's Bay the morning of Tuesday, 3 February 1931 dawned like any other mid-summer day, a little still and sultry, but warm, fine and with a promise of a sleepy afternoon and long balmy evening to follow. The sea was spray-swept but calm, a contrast to the 'terribly rough' conditions of the previous two days. Turbulent waters had thrown 'vast quantities of queer seaweed on the beach' at Waimarama, where local resident Dorothy Campbell thought the sea 'looked a most peculiar colour, so much so that everyone was talking about it, saying that there must either have been some sort of submarine disturbance or a storm at sea'.[4]

As the sun climbed, people perhaps started their day with porridge, eggs, toast and a cup of tea. Some men and women — more men than women in those days — left their flats, bungalows and villas to go to work. Others left in the hope of finding any kind of miserable job to keep the wolf from the door. This was the second summer of the depression. Unemployment had soared to levels not exceeded until the 1990s, and many families faced unprecedented hardship.[5]

In some households the day began with a turfing of sleepy children from their beds. It was the first day of term after the summer holidays. Mothers across Hawke's Bay packed reluctant youngsters off before beginning their own day's chores, perhaps boiling the copper for the laundry or getting out the carbolic soap to give the floors another scrubbing.

For Napier harbourmaster Captain H. White Parsons the day started with an unexpected surprise. He had been expecting the sloop HMS *Veronica* with her crew of 104 to arrive that afternoon,[6] but was told she would arrive early at 7.00 a.m. The warship was safely berthed in the inner harbour before eight. Her captain, Commander H.L. Morgan, DSO, prepared for a round of official visits.[7]

In Napier Hospital, Dr T.M. Gilray prepared to perform a routine appendectomy.[8] Havelock North resident Mary McLean left 'the village' — as the close-knit community was nicknamed — for a fourteen-mile (23-km) trip to her Napier bank. Her cousin, 71-year-old elder settler Bernard Chambers, left his new Te Mata homestead for Havelock North to buy parts for his household water supply.[9] Molly Donnelly and Jack Chambers prepared for their wedding, scheduled for that morning in St Luke's Church. On Tutira station, erudite naturalist-farmer Herbert Guthrie-Smith set off with a station hand to muster a block north of the little lake.

None of the 70,000-odd people of Hawke's Bay — the towns Napier and Hastings, the villages of Havelock North, Taradale and Clive, the orchardists and horticulturalists of the Heretaunga plains, or the farmers and sawmillers

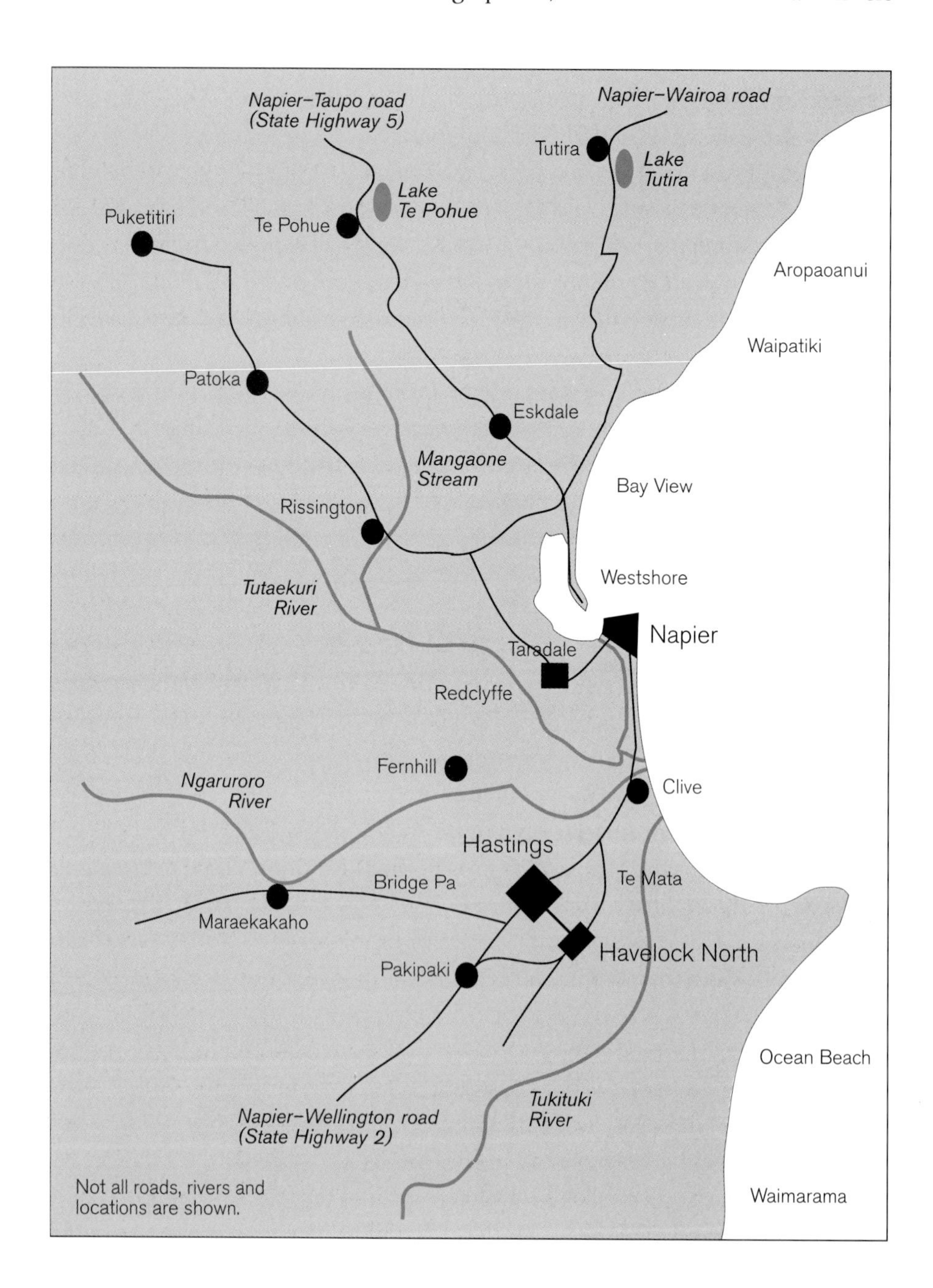

Hawke's Bay, February 1931.

of the country beyond — knew of the titanic stresses building up under Aropaoanui, a narrow knife-slash of shingle amid the coastal cliffs eight miles north of Napier. Years earlier, rock surfaces nineteen miles (31 km) underground had jammed together, slewing the remorseless advance of the two great crustal masses underpinning Hawke's Bay like an off-centre brake, creating tensions that could only be released by cataclysm.[10]

Earthquake — 10.46:47 a.m., 3 February 1931

A few people felt uneasy as the morning wore on. Havelock North resident Wilf Leicester, recovering at home from a broken leg, noticed that the air 'became very still and there wasn't a sound, not even a bird singing'.[11] At Waimarama, Dorothy Campbell saw that the sea was now 'so calm & still that Bro. Frank ... remarked on it'. The air 'had grown still & oppressive'. She felt so uneasy inside her house that she suddenly:

> ... jumped up & ran to the kitchen where Ann, nurse & the maids were having morning tea. I don't know what made me, or what I said, or what I did & neither do they, but I called something & the others were out & I was just out as the earthquake hit the house.[12]

For others there was no warning. At 10.46 a.m. they were going about their ordinary business. Just over 46 seconds later the earth staggered before energies roughly equal to the detonation of 100 million tonnes of TNT.[13] Much of this was channelled down a moving slab of landscape — a 'rupture zone' — that ran directly below Napier and southwest across the Heretaunga plains.[14]

Most people felt the first sledgehammer blows as an uplift. Dogs howled, cats ran screeching, and horses — still hauling suburban milk carts and trade wagons in 1931 — reared and tried to bolt. All went unheard amid a 'tremendous noise' that Llewellyn Mitchell des Landes, working in the meter repair shop of the Napier Gas Company, compared to 'an express train'.[15] Buildings lurched violently, many shedding outer walls or decorative pediments. People inside were hurled this way and that, some injured by furniture and debris, or pinned by collapsing ceilings and roofs. Others, caught on footpaths, were injured or killed by debris crashing from walls and buildings. Chimneys in the housing districts bent like reeds in a gale, then cracked and broke, sending debris tumbling. Telephone and lighting poles swayed abruptly, some remaining canted at crazy angles. Vehicles skittered on roads as the carriageway surged and rippled.

About 30 seconds passed. Suddenly the ground heaved again, a different kind of movement that some felt as a downwards jolt. This time the effect was completely devastating, mind-numbing waves of destruction that swept across the province, smashing weakened buildings and walls. Rubble poured into the streets, and many who had rushed outside after the primary shock died as shattered masonry crashed on and around them. Avalanches of bricks and debris slammed into vehicles, a few with their occupants still inside. The tortured earth rumbled, a massive sound punctuated by the crisp treble of shattering glass, the bullet-like cracking of buildings, the thuds and thumps of falling furniture, the crash of glassware and crockery, and the sliding rush of collapsing masonry.

At last the shock waves rolled by, leaving a terrible trembling in the ground that some observers compared to boiling water. For a few moments afterwards the silence seemed complete. Dust from crumbled mortar and shattered concrete in the business districts of Napier and Hastings rose into the air, thickening as the moments passed to a white, thick, powdery fog that briefly obscured vision even a block away.

Wrecked buildings on Napier's Hastings Street a day or two after the disaster. Fire completed the devastation begun by the quake. (PHOTOGRAPHER UNKNOWN, T. PHILLIPS COLLECTION, ALEXANDER TURNBULL LIBRARY, F-57112-1/2)

Havelock North resident W.H. Ashcroft was working in Napier when the earthquake struck.

'The earthquake continued for about 2 1/2 minutes, during this time we could only hold on to one another and wait for what would come next. We were afraid that our wall and Blythes would come down, but we could do nothing; the wall of the Ford Garage, a new building, took on the most extraordinary contortions, a convulsion would come and the wall would wriggle from the bottom to the top like a snake, sometimes it would bend over and very nearly hit the Post Office. All the time there was the most awful racket of falling buildings and a fearsome black dust so thick that one could not see. The peculiar thing was that I felt all this was happening to others not to me, and I was merely a spectator; others have told me since that they had exactly the same feeling. I think, as the quake continued, we all felt we were certain to be killed and just wondered how soon it would be; again others have said they felt just the same. After what seemed an interminable time, the violence of the shocks ceased, although the ground was trembling all the time ...

'... it was not till we were in Dickens Street that we realised the magnitude of the disaster. Dickens Street was a mass of bricks and mortar; nearly every building was down. Hastings St. looking towards Shakespeare Road was impassable. I got as many as I could to an open space behind Simmonds & Co. and we stayed there as further shocks came and we had to hang on to one another to keep from being thrown down ...'

W.H. Ashcroft[16]

The view south along Hastings Street, some time after the quake. The Union Bank ruin is visible on the right.
(Photographer unknown, Alexander Turnbull Library, PAColl-7270, F-2945-1/2)

NAPIER EXPERIENCES

Everyone who lived through it experienced the earthquake differently. W.H. Ashcroft was sitting in his office in Napier's business district when he felt the uplift, which he compared to a terrier shaking a jack rabbit. His office fell apart around him; he looked up and saw blue sky.[17] Eighteen-year-old Jessie Atkinson, staying in a house on Napier Terrace near the hospital, 'watched a piano lurch from one side of the room to the other and back again'.[18] Shakespeare Road store owner H.R. Hounsell, who had been caught by the 1906 San Francisco quake, hastened to pull his wife to safety as the floor heaved, bruising his legs as he tried to carry her from the shop.[19]

In the Napier Technical School on Munro Street, not far from the railway station, Miss M. Bampton's second-year commercial class of 35 had just reconvened after morning play when the earthquake shattered the building. Bricks showered into the room. Dazed and with blood covering her dress, Bampton told the frightened children to make their way outside. She crossed a plank over a collapsed section of the upper floor, and was helped down the stairs.[20] Her class escaped — she thought during the interval between the first and second shocks — but others did not. Fellow teacher W. Olphert yelled at the boys to dive under their desks as the earthquake slammed the building; however, 'several dashed into a narrow corridor, where they were buried under; the fallen walls'. Other boys escaped, then 'rushed back to extricate their pals, and at least two got no medals for their bravery, but were shockingly injured when the remainder of the building fell'. Olphert was intensely admiring. 'Truly one can be wonderfully proud of these boys. They may not have much knowledge, but they have something infinitely more precious.'[21]

Building after building swayed in the town centre, sending debris pouring into the streets as curtain walls and facades failed. The Masonic Hotel at the seaward end of Emerson Street cracked and broke. One occupant was 'near the kitchen when the shake came'. As 'huge blocks of masonry' fell about her, she ran for the nearest door where she found another woman. They could get no further and crouched as the 'building crashed about our ears'. Choking on the plaster dust, they pushed their way out through 'heaving masses of debris'.[22]

Local tobacco magnate and philanthropist Gerhard Husheer was recovering from hip surgery in Dr Moore's private hospital on the Marine Parade. The bed in which he was immobilised skittered across the floor while a wardrobe rocked crazily and crashed down. Suddenly the three-storey building began to tilt backwards. Husheer feared for his life as his bed slid to the back of the room, but when the angle reached about fifteen degrees the collapse stopped. The hospital had slumped into an underground carpark.

The teetering Marine Parade frontage of Dr Moore's hospital remains one of the best known images of the disaster. It 'tilted back several feet in the front to lie at a queer angle' and almost became the tomb of tobacco magnate Gerhard Husheer. Ironically, the building slumped because of an ill-advised excavation to make an underground carpark.
(STORKEY COLLECTION)

Looking south along Napier's Marine Parade. Smoke still rises from the devastated town in this image taken the day after the quake. Dr Moore's hospital tilts centre-frame.
(JOHNSON COLLECTION)

Rubble slumps across the waterfront in this picture of the Napier town beach, looking north. The last debris was not cleared until the 1960s.
(JOHNSON COLLECTION)

Visiting doctor H. Campbell-Begg was with a patient in an upper-storey consulting room on Hastings Street when the earthquake turned the rooms into a confused mass of burst cupboards and spilled medicines, jumbled in 'one indescribable litter on the floor'. The second downwards shake told him it was time to leave.

> The stairs are still intact, the whole earth is boiling. That feeling of bubbling, gentle, vertical, up and down, vertiginous motion which follows great shakes, defies description. The street is reached. Hastings Street — a rubble-filled lane. The fallen buildings shelve on either side in ramps of debris in the middle. Power lines and telephone wires lie in confusion on the mass. The air is filled with haze and dust — not a dense, choking fog of dust, but one which gradually thickens until nothing can be seen through it more than a block away ... The sounds of nature are stilled. That thunderous roar, when everything came rumbling down, has exhausted itself.[23]

Linotype machines toppled in the *Daily Telegraph* offices and the 'operators had an anxious time dodging the molten metal which was splaying from the open melting pots'. Compositor apprentice Val Harrison was trapped by a falling wall and killed, but was the only fatality there. The quake hurled the whole front upper wall of the building into the street.[24]

The wrecked BNZ building on Hastings Street before the fire. Three days after the disaster, the *New Zealand Herald* reported that 'hazardous work remains to be done in the removal of standing walls, some of which are left at dangerous angles'.
(STORKEY COLLECTION)

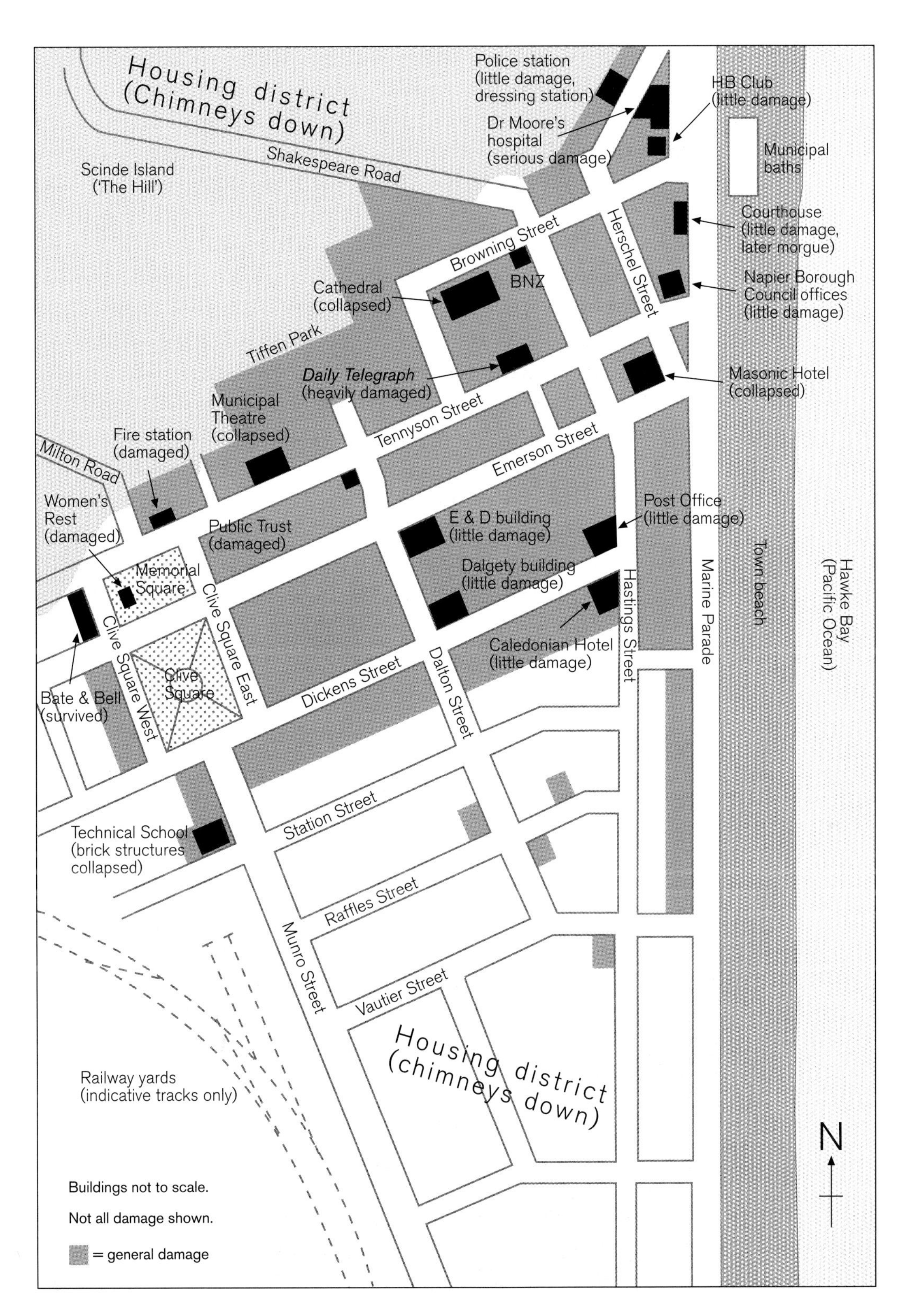

Quake damage, Napier town centre, 3 February 1931.

Source: 'Council of Fire and Accident Associations of New Zealand. Official records of Napier Earthquake, February 3rd–10th 1931'. Also sources cited in text.

People staggered into the streets. One woman was dragged clear of the Union Company building only to be killed by a piece of falling masonry. In the Salvation Army maternity home, Nurse Thomson felt the earthquake and rushed to save nine babies who had been taken out into the sun earlier. She had almost reached the infants when a chimney collapsed among them. By what she called 'an act of Providence' they were uninjured, and the six other babies in the home, with their young mothers, also survived unscathed.[25]

Napier South, the triangle of suburb beyond the town centre, trembled before the full force of the quake, catching Jean Anderson in the middle of writing to a friend in Wanganui. Ink splashed her page before she was slammed to the floor.[26] The pocket-triangle of suburban quarter-acre sections had been established 20-odd years earlier on barely compacted silt. Those in Nelson Park, the huge sports oval in the middle of the suburb, heard the roar and looked up to see Bluff Hill collapsing, engulfing the town within seconds in a cloud of dust. Then the turf began 'heaving and rolling like a foundering ship'.[27] Shock waves rolled over the park, clearly visible in the grass. Water gushed high into the air, squeezed from the aquifer beneath.[28] Houses groaned and creaked. A few bounced from their piles or sagged as the ground gave way; W.J.C. (Bill) Ashcroft later saw houses in Georges Drive 'lying at all angles where the foundations had given'.[29] Chimneys crashed down all over town, causing casualties in the suburban areas and killing three-year-old John Canham, who

Opened in 1912 after a ferocious local political debate, and at a cost of nearly £27,000 (equivalent to about $1,944,000 in late-twentieth-century money, taking inflation into account), Napier's Municipal Theatre could seat more than 1000 people. Its massive stage was reputed to be one of the largest in Australasia. This is all that was left after the quake. (STORKEY COLLECTION)

Napier's ruined and burnt Masonic Hotel, thought to be one epicentre of the fire that tore through Hastings Street. 'No-one could be alive under those ruins,' one witness declared of the townscape the day after the quake. 'The bricks lying on the pavement were so hot that they burned the soles of my shoes. Smouldering timbers and portions of walls were falling all the time.'
(Hare Collection)

Naval ratings amid the rubble in Napier's Emerson Street.
(Photographer unknown, Alexander Turnbull Library, F-121983-1/2)

was playing in his front garden.[30] Other chimneys, weakened by the shock, fell in the minutes after the quake.[31]

At Napier Boys High School, in the southern corner of the town, the older pupils were practising military drill outside; new starters were in class and rushed out as the quake hit. Headmaster W.T. Foster was almost crushed by masonry collapsing from the assembly hall — a falling half-tonne fragment 'broke in two and when the dust cleared he had merely a bruised leg'.[32]

Friable limestone cracked and slumped on the hill, scalloping Bluff Hill and the Faraday Street valley, leaving some houses teetering on the brink. Mrs L.T. Bisson and her washerwoman rushed outside as the quake rattled their house, but 'no sooner had they left the house than the whole retaining wall fell, killing both'.[33] Mary Hunter tried to leave her Fitzroy Road house, making four attempts to climb the steps to the road above. When she finally reached the top she saw 'hundreds of spiral yellow dust clouds going up in the air from the fallen chimneys of every house in sight'.[34] At Napier Girls High School, on the hill above the town, an exit door jammed but 'panic was stopped' when one pupil held the 'door of the reeling building open while her schoolmates escaped'.[35] In nearby Corry Avenue, teenager Eileen Cray, with 'great bravery and coolness of mind', ran into a swaying house and climbed a 'dangerous staircase amid falling brickwork and timber' to rescue a baby.[36]

The shock smashed into the public hospital, a long-standing landmark at the southwestern end of the hill and by 1931 the only base hospital in the district.[37] It levelled the isolation block and wrenched ward after ward,

Slips scalloped large sections of the Napier hill, throwing up choking clouds of dust and leaving houses standing precariously on the brink. These slips occurred above Carlyle Street. In 1939, J.G. Wilson suggested that no quake like the 1931 event could have occurred in Napier for a thousand years because there was no evidence of earlier slips of this magnitude.
(ALEXANDER TURNBULL LIBRARY F-60957-1/2)

smashing some of the buildings entirely and leaving others shattered and twisted. The adjacent Spanish Mission nurses' home swayed violently and disintegrated, watched by horrified local resident Criton Smith, who felt a 'terrific jolt' and heard 'screaming voices' before the home 'collapsed like a house of cards', amid more screams and cries. She clapped her hands to her ears to block out the sound and saw one nurse run from the doorway 'as a heavy piece of masonry crashed behind her'.[38]

Others had a more distant view. For K.C. Sinclair, watching from a nearby Balquidder Road property, the nurses' home simply crumbled with an appalling noise. 'From the dense cloud of dust I imagined at first that it was an eruption,'[39] she recalled. Jessie Atkinson saw only the aftermath. Picking herself up from where she had been thrown, she looked out of a window to see that where the 'lovely architecture' of the home had 'stood against the very blue sky was a column of dust rising higher and higher'.[40] Some later compared the destruction of the home to the effects of shellfire.[41]

There were nineteen people inside, including sixteen nurses asleep after the night shift.[42] Seven nurses died immediately, among them 24-year-old Nancy Thorne-George and 19-year-old Eileen Williams.[43] An eighth nurse succumbed later. Half a dozen nurses were extracted with serious injuries, and only three escaped uninjured. Napier Terrace resident George Brown almost broke down when he saw the rubble soon afterwards.[44] C.E. MacMillan, the Tauranga MP who visited the site a few days later, saw nurses' personal possessions lying in a heap near the wreckage and was saddened to realise that they had been killed.[45]

Napier's Spanish Mission nurses' home was a triumph of style when it opened on the corner of Chaucer Road and Napier Terrace in 1930, and an important new facility at a time when money for public health services was at a premium. (Storkey Collection)

The pathetic remains of the nurses' home. Faulty design and construction, compounded by large openings in the lower floor, contributed to a 'frightful … toll of life'. Eight nurses died, along with three office staff who, on the bottom floor, were 'in a hopeless position'. (Storkey Collection)

Few wooden houses completely collapsed during the quake. Those that fell down usually did so for a specific reason – in this example, above Napier's Faraday Street, the ground beneath the foundations failed. The Carlyle Street slips are visible to the right.
(Storkey Collection)

Other slips crashed down from Bluff Hill, the sheer northeastern end of the Napier hill. Fears that the rubble had entombed a car with its occupants were not allayed until the last of the debris was cleared in the 1960s.
(K.S. Williams Collection, PAColl-1960-10, Alexander Turnbull Library)

AHURIRI AND THE ROADSTEAD

George Brown was in Ahuriri when the earthquake struck and 'suddenly thrown full length on the floor ... there was no possible chance of regaining my feet'.[46] Not far away, harbourmaster H. White Parsons had boarded HMS *Veronica* to see Captain Morgan. They were sitting in Morgan's cabin when 'suddenly we heard a terrific explosion'. Both men thought the magazines had blown up, but when they rushed on deck they found a greater disaster in progress.

> The corrugated iron walls of the stores on the wharf were bursting asunder and disgorging bales of wool. The railway lines were twisting and bending under our eyes, and, with a crashing sound, the wharf a few yards in front of us gave way and fell into the harbour. The bed of the sea rose beneath us, and the stern wires gave way.[47]

The harbour bottom slammed into the keel and five of the six heavy hawsers holding the sloop to the quay snapped. The quayside astern — where Morgan had originally intended to moor — slumped and broke.[48]

Out in the roadstead, the freighter *Northumberland* had been at anchor for some days taking on frozen meat ferried out by lighters, an expedient that reflected 60 years of bitter debate about Napier port services. The earthquake shook the ship violently, and eighteen-year-old cadet A.F.R. Irwin and others on deck saw an old wreck boil to the surface. The hulk paused long enough for Irwin and others on deck to read the name *Northumberland* on the stern — then vanished again 'with a loud sucking noise'.[49] This was no spectral warning of doom but the remains of an earlier *Northumberland*, a full-rigged iron freighter wrecked off Napier in May 1887.[50]

On board the motor vessel *Taranaki*, anchored in the roadstead nearby, radio operator J.J. Grundy was writing a letter when the boat shook so violently that his pen 'jumped right across and off' his writing paper. He was swung around in his chair, found himself facing the clock, and jotted down the moment in the communications log. The freighter vibrated as if in a gale, her derricks and rigging sounding like harp-strings. A few on board thought she had blown up, then someone cried, 'Look at the shore!' Horrified seamen and dockers saw 'the town of Napier crumbling before their eyes amid a fog of dust'. Grundy hastened on deck to watch as 'the whole of Napier' seemed to 'elevate itself and then subside'.[51] Bluff Hill slumped. All around the coast they saw 'great clouds of dust from falling headlands', as if the shore was under bombardment. The spit 'undulated for its entire length' and one of the harbour beacons 'waggled like a pendulum' before settling on a 30-degree angle.[52]

Suddenly the sea bottom smashed into *Taranaki*'s keel — no slight touch but a massive blow that observers half a mile distant on shore heard as 'an ominous grating noise'. The engineer feared the ship would 'surely break her back or suffer some structural damage'.[53] The ship floated off as the quake subsided. Soundings revealed that the freighters now lay in just four fathoms (a little over seven metres) of water, barely enough to carry them. It had been seven fathoms deep when they had anchored. Half an hour later the SS *Waipiata* arrived, reporting shoaling in the roadstead and only three fathoms of water outside the harbour.

HMS *Veronica* at the west quay. Her mooring cables snapped during the quake – only the flax-fibre ropes held – while 'buildings crashed down all along the waterfront' and the ship was left resting on the bottom. Crew feared she might cant over if the last lines broke. Collapsed wharfing gives graphic evidence of the violence of the event. Commander Morgan originally thought of mooring at about the point where this picture was taken; had he done so, HMS *Veronica* would have been hit by the falling wharf and almost certainly suffered heavy damage. (P.T.W. ASHCROFT COLLECTION, ALEXANDER TURNBULL LIBRARY 139887-1/2)

MV *Taranaki* was one of three merchants in the vicinity when the quake struck. Hammered by the rising sea bottom, she and the SS *Northumberland* ploughed out to sea afterwards in search of deeper water. This picture was taken in 1942. (RNZAF OFFICIAL, ALEXANDER TURNBULL LIBRARY C-24715-1/2)

GREENMEADOWS AND TARADALE

Thirty-two people died in Greenmeadows and Taradale townships, about five miles (8 km) southwest of Napier. The Marist Brothers seminary at Greenmeadows was heavily damaged. Thirty-four students and four priests were in the chapel on the second day of a retreat, and as the building swayed they hastened for the door and the stone vestibule that linked the chapel with the seminary. Most were caught by collapsing masonry. Seven students were killed instantly. The rest sheltered under their seats and climbed to safety over the rubble when the earthquake stopped. Three could not be accounted for, but when 'hope was almost abandoned', searchers heard groans and found one seriously injured student.[54]

The remains of the chapel at the Mount Saint Mary's Seminary, Greenmeadows, where several students along with Fathers Gondringer and Boyle were trapped by 'an immense fall of stonework'. For the casualties there was 'no opportunity for a funeral or a Requiem Mass'. They were instead buried at the Puketapu Cemetery in graves dug by fellow students.
(KITTY WOOD COLLECTION, PACOLL-1009-13, ALEXANDER TURNBULL LIBRARY)

The fallen statue of the Blessed Virgin at the Mount Saint Mary's Seminary, Greenmeadows.
(KITTY WOOD COLLECTION, PACOLL-1009-04, ALEXANDER TURNBULL LIBRARY)

Ninety-year-old James Collins, who had served in colonial-era India and later worked for Hastings pastoralist Major-General A.H. Russell, was rescued from the ruins of the Park Island Old People's Home after three days. Trapped by a heavy beam, with his face in a kapok mattress, he never lost hope. 'I would call until I was exhausted and then sleep,' he told reporters afterwards. 'Then I would call again.' On the sixth attempt he was answered. Fourteen of the 94 people in the home died, among them 73-year-old John McKenzie, 72-year-old John Rae and 66-year-old John Thomas.
(Storkey Collection)

The destruction in Hastings was almost as thorough as in Napier. 'It seemed that even time itself had been obliterated by the upheaval,' a reporter later wrote, 'as for one horrified moment the whole town was wrapped in a death-like silence.' James Daroux took this picture of the wooden shops at 104–114 Heretaunga Street West for insurance purposes after the quake.
(J.H. Daroux Collection, PA1-f-145-42-40, Alexander Turnbull Library)

In the independent borough of Taradale, the town hall, library and 'Ladies' Rest' were destroyed, along with the public toilets, band rotunda, Masonic Hall, the Taradale hotel, the police station, seven shops and two houses. The Soldiers' Memorial gained a drunken tilt.[55]

Power for the whole region came from a transformer station at Redclyffe, just outside Taradale, where 110-kV national grid power was stepped down to 11,000 volts for distribution to local towns and districts via the regional ring main. Hooking Hawke's Bay to the national grid had been one of the triumphs of the 1920s. By 1931 the whole region relied on it, with the exception of Havelock North, which had its own hydro scheme. The Redclyffe Power House was of monolithic concrete construction and survived the quake; however, the lead-acid batteries that normally powered the oil switches were toppled, and the transformer mountings failed. Three of the heavy devices 'rolled off their pads and crashed to the ground', and all lost their oil.[56] Power to the 11,000-volt regional ring main was immediately cut off, with no hope of quick restoration. This had its good side — those caught among falling power lines were not electrocuted — but ultimately it sealed the fate of Napier's town centre. Power was unavailable when desperately needed to pump fire-fighting water.

Incoming 110-kV grid power was also cut off. A landslide swept away the foundations of a transmission tower near Pihanui, forty-odd miles (60 km) north of Napier. The pylon collapsed, taking with it the 110-kV line from Tuai. The undamaged southern line to Mangahao, near Shannon, also went dead when short circuits opened the primary circuit breakers. This happened a full minute before the earthquake was felt there — it was strong enough to fell chimneys.[57] The breakers were reset within fifteen minutes, and by 12.45 p.m. the engineers had brought the 5000-kW Wellington steam plant into operation to make good losses from Tuai — a move possible because the national grid was deliberately designed from the outset as a single integrated system. However, the damage to the Redclyffe transformer station prevented this power being distributed to the Hawke's Bay regional ring main.[58]

DEATH IN 'THE JEWEL OF THE PLAINS'[59]

The quake hammered Hastings — the region's primary rural service centre — with horrific force. 'Hastings was full of shoppers when the disaster occurred,' a *New Zealand Herald* reporter wrote two days later, 'and they were caught like rats in a trap.'[60] In one 'mighty upheaval' the town became 'a vast charnel house', the main business street reduced to a 'gully of destruction beneath whose ruins could be heard the cries of women and children'.[61] Buildings swayed while pediments crashed to the pavement and interior fittings slid and rolled. In Mayor G.F. Roach's drapery store on the corner of Heretaunga and King Streets,

ABOVE: The greatest single tragedy in Hastings happened in Roach's store on the corner of Heretaunga and King streets. There were 50 people on the top floor alone when the earthquake struck. Seventeen died as the building collapsed.
(STEFFANO WEBB COLLECTION, PACOLL-3061, ALEXANDER TURNBULL LIBRARY G-19263-1/1)

LEFT: The collapsed bell tower had been cleaned up by the time James Daroux took this picture of the Hastings Post Office for insurance purposes.
(J.H. DAROUX COLLECTION, ALEXANDER TURNBULL LIBRARY C-23673-1/2)

employee Vera Smith saw the 'front of the building' rise up in the air before the building cracked and broke.[62] There was time for one of the assistants to rush into the arms of Christchurch visitor Ernie Weston. He tried to calm her, when suddenly:

> ... one of the steel pillars fell right across the girl's body, and she was crushed alongside the counter. She pleaded for freedom, but the big pillar held her in such a position that in a moment her life was gone.[63]

Weston had 'missed death by inches', sadly reflecting later that 'many of the staff never had a chance'.[64] A few escaped. Shop assistant A. Sampson was 'serving a lady who had a baby in a motor car outside', when the building and counter began to rock.

> 'My God! it's an earthquake,' shouted the customer, 'get outside quick.' I was dazed for a second and the next thing I knew was that I was being dragged out into the street by the customer. Buildings were beginning to fall all round us, and I looked through the doorway where I had escaped and saw two girls trying to get out. They were seriously hurt, but still struggling. I tried to get back to help them, but the customer who had rescued me held on to me. The next second a huge crash was heard, and then the debris completely enveloped my friends. Oh! I shall never forget the look on their faces.[65]

Wrecked buildings along King Street, Hastings.
(PHOTOGRAPHER UNKNOWN, ALEXANDER TURNBULL LIBRARY, F-20248-1/2)

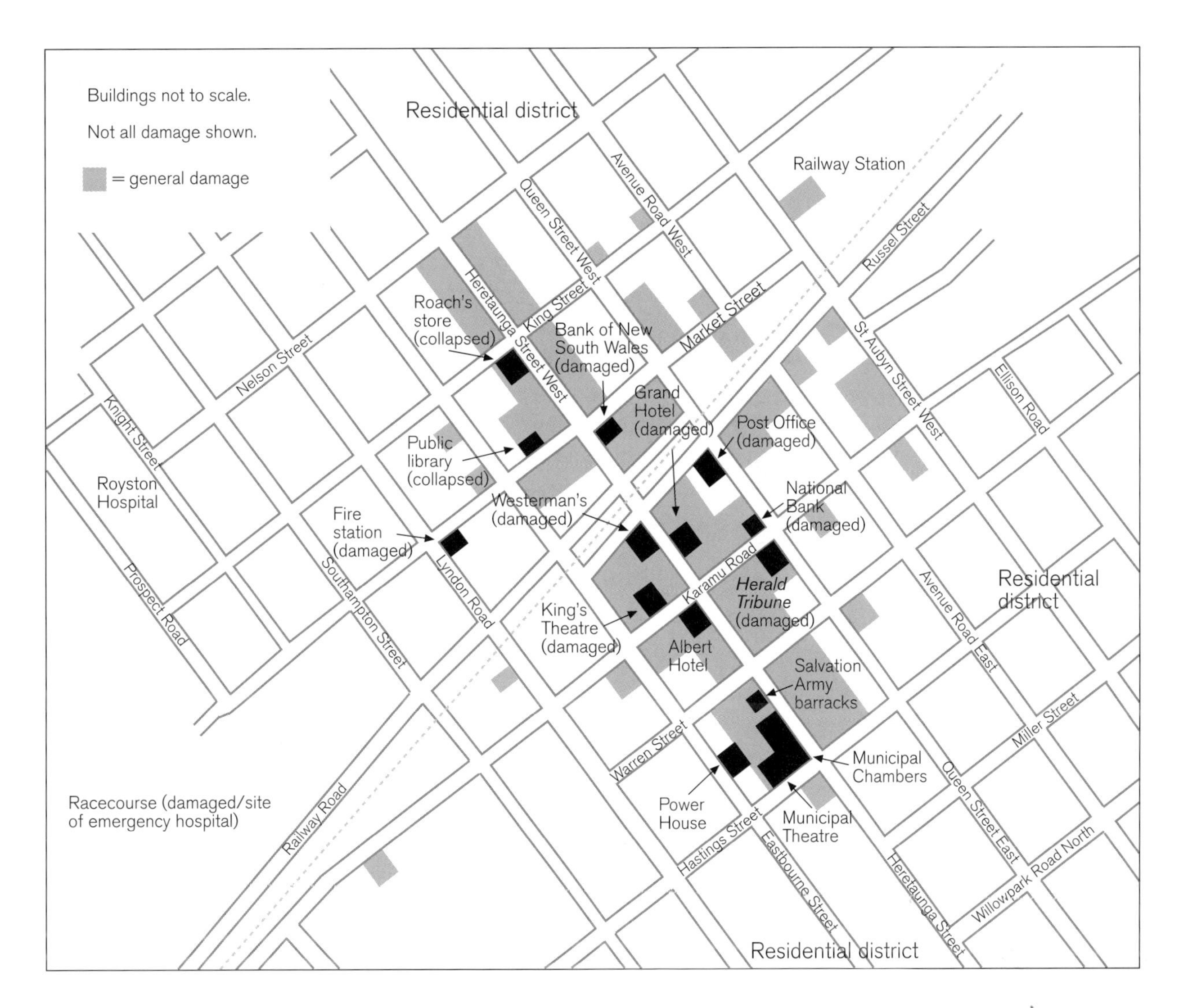

Quake damage, Hastings town centre, 3 February 1931.
SOURCE: 'COUNCIL OF FIRE AND ACCIDENT ASSOCIATIONS OF NEW ZEALAND. OFFICIAL RECORDS OF NAPIER EARTHQUAKE, FEBRUARY 3RD–10TH 1931'. ALSO SOURCES CITED IN TEXT.

There were more than 50 people inside the building when it fell. Some escaped unhurt. Vera Smith was buried under a showcase but managed to scramble out. 'Everywhere, people were crawling out of the rubble,' she recalled.[66] Others were not so lucky. Weston himself looked around to see 'daylight shooting through the debris'. He heard appeals for help 'coming from everywhere' and saw, 'in what was once the boot department', the figure of a man desperately fighting for his life. He was beyond help. Fire erupted in one corner of the building. Weston hastened to help others lift 'a heavy beam from across the shoulders of Mr Roach jun[ior]. He was lucky to crawl out alive.'[67] Seventeen people died in the store, including a boy who had come to get a new school cap.

Further down the road, *Herald Tribune* reporter A.L. Ryan died when the Post Office clock tower fell on him — he was still listed as missing two days later.[68]

There was almost a calamity in the *Herald Tribune* building on the corner of Karamu Road and Queen Street East. Reginald Gardiner stopped a typist running from the doorway just as the parapet crashed down. 'All the rest of the staff miraculously escaped,' a reporter later wrote, 'except Mr Bluett, a linotypist, who was badly injured.' It was remarkable that there were not more casualties among typesetters at a time when the art was fairly described by the term 'hot lead'. Reservoirs of molten metal were easily spilled when heavy linotype machines were toppled by the quake.
(PACOLL-3034, ALEXANDER TURNBULL LIBRARY F-4868-1/2)

In the *Tribune* offices on the corner of Karamu Road and Queen Street East there was a general rush to leave the upper floor. Reginald Gardiner stopped one of the typists running outside moments before a concrete parapet crashed over the doorway. The Grand Hotel on Heretaunga Street swayed and partly collapsed, hurling a porter from the upper floor into the street, killing eight people and trapping proprietor J.A. Ross in the cellar. He died that night when fire swept through the ruined structure.

Not far away the Cosy Theatre — a popular movie venue capable of holding 1000 people — fell 'as though its foundations had been swept from underneath it'. The public library on Market Street crumbled into a 'hopeless tangle' of debris that engulfed both staff and readers.[69] Auctioneer Tom Gill was buried in debris while trying to carry an unconscious woman to safety.[70] Two women were 'thrown completely under a car', where they were seriously injured.[71] Few of those hit by the debris survived unscathed. Among many others, Betty Percy received head injuries, Dorothy Rust head injuries and a broken leg, and J. Sheffield Fergusson an injured hip.[72]

The earthquake hit nearby Havelock North with scarcely less force. For three decades the people of this tiny settlement, nestled against the hills on the edge of the Heretaunga plains, had regarded their town as 'our village', a special community where all knew each other and shared common interests in literature, the arts, spiritualism, education and music. Up to a third of the thousand-odd residents in 1931 reputedly belonged to a hermetic religious order.[73]

Beekeeper Bill Ashcroft 'was just coming around the house … when there was a rumbling sound and the car in the garage started to shake and shiver as if the engine were running very badly'. He dragged his mother from the house and the two stood 'hanging on to the clothes line watching the house shake while the ground heaved and bumped beneath us'.[74] Bernard Chambers, back at his Te Mata home after his morning drive into Havelock North, was about to make a phone call when the quake struck 'in a second'. His nine-year-old brick house was not quake-proofed, and in two more seconds the south wall of his office fell out, leaving 'a place big enough for a car'.

> [I] tried to get … outside but [the door was] blocked & [I] managed to get under some fallen bricks in [the] passage & out by [the] front door, the house danced and rocked continually & to see the buildings crumbling was a terrifying sight.

Quake in progress; this picture of the wooden bridge over the Karamu Stream, between Hastings and Havelock North, was apparently taken as the shock waves rolled past. The bridge collapsed, taking with it the water mains to Hastings.
(HNL 3041, Havelock North Public Library)

There was considerable devastation in the Havelock North town centre. Apart from Foster Brooks building, visible in the background, there was damage to the Town Board offices – where brick walls collapsed – the Weathered building, Fryer's buildings and the tower of St Luke's Church.
(S.C. Smith Collection, PAColl-3082, Alexander Turnbull Library G-48567-1/2)

A closer view of Foster Brooks building on the corner of Te Mata and Joll Roads in Havelock North. It was one of the few structures totally lost in 'the village'.
(HNL 670, Havelock North Public Library)

There were no casualties in Havelock North itself, but parts of 'the village' suffered significant damage even so. Here are the Estaugh & Treneman premises after the quake, photographed by James Daroux for insurance purposes.
(J.H. Daroux Collection, PA1-f-145-65-6, Alexander Turnbull Library)

His daughter Hazel Nairn and her sister-in-law Beatrice ran clear of the front porch, while his wife Lizzie 'ran downstairs through the drawing room and main verandah'. Chambers added 'all unhurt'.[75]

Gwen Moran, a pupil at Havelock North Primary School, was on her way with Standard 1 and Standard 6 pupils to the swimming pool, accompanied by headmaster Henry Lyall. The line halted between the bathing sheds and one of the pools, and at that instant the earthquake struck, sluicing the children with water. 'I hear a voice shouting "It's the end of the world — the end of the world", adding momentum to the panic stricken screams of sheer terror.'[76] The temporary caretaker seized the torn ends of the lighting cables and threw them clear.

At Waimarama, a wide stretch of golden sandy beach on the coast, Dorothy Campbell had already given a warning before the earthquake hit, and with her family had:

> … managed to reach the fence before we were thrown on the ground & there we were with it heaving up & down like the waves of the sea & roaring & crashing & banging, so much so that one literally could NOT hear the chimneys come down or the crockery breaking. Almost as I had been knocked over I pulled myself up again by the fence to see how everyone was. Nurse was sitting with Ann, who was repeating over and over 'never mind Mummy, never mind, never mind mama' … the two maids were holding each other & trying to get up. All this takes ages to tell but it was literally only seconds. We were facing the sea & I saw an island jump, as I thought, about fifteen feet out … & at the same time a reef of rocks which I had never seen before appear between the island & the mainland as well as other rocks appear.[77]

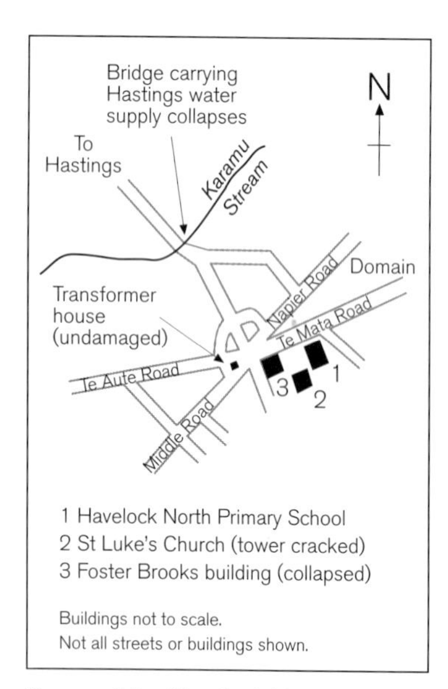

Some of the Havelock North earthquake damage.

HAWKE'S BAY SHATTERS

The shock waves spilled across Hawke's Bay. Near Tutira, about as far northwest of the epicentre as Napier was south of it, Guthrie-Smith and his shepherd were 'moving sheep across alluvial lands at the north of the lake' when:

> Without any sort of warning or premonitory sound whatsoever the ground began to shiver and tremble, then seemed to simmer as water coils and convolutes ere bubbling to the boil. This was instantly followed by jerkings and jolts violent enough to stagger and then to throw us down, but not to roll us clean over. By the use of hands and arms it was possible to maintain some sort of sedentary equilibrium.[78]

Photographers captured dramatic images of vehicles tipped into chasms, particularly on the Embankment between Napier and Westshore. Most of the vehicles damaged in the quake were smashed by falling debris in the Napier and Hastings town centres. Even these numbers were relatively small; a 1990s assessment of vehicle losses from the quake concluded that almost all the regional stock survived undamaged.
(TOP: JOHNSON COLLECTION) (BOTTOM: PHOTOGRAPHER UNKNOWN, D.H. JONES COLLECTION, ALEXANDER TURNBULL LIBRARY, F-135767-1/2)

The quake damaged the foundations of the new Mohaka Viaduct, then going up to take the Napier–Wairoa railway line across the deep Mohaka Gorge. One man died when a caisson collapsed, and another was half buried in the rubble. This picture was taken just a few days before the earthquake.
(PA1-0-643-092-2, Alexander Turnbull Library)

Further north again, at Mohaka, there was a sound like artillery fire. Standard 1 pupil Darry McCarthy felt the ground leap 'like an unbroken horse ... over and over, leaving great fissures around us'. She watched a huge slab of land — some 300 acres (122 ha) — collapse from Mohaka Station into the sea.[79] The noise was tremendous. Dust rose around the landscape, making it impossible to see. The violence of the quake became obvious later, when a snapper was found beached 50 feet (15 m) above sea level, where the sea had splashed under the impact of the collapsing coastline. The Mohaka hotel and an adjacent house belonging to J. O'Grady caught fire and were destroyed. The teacher and children at the Mohaka school were able to escape even though cupboards fell across the doorway and obstructed their exit.[80] Guy Gaddum's house virtually collapsed; only the walls were left standing, and J.R. Murphy's home slipped from its foundations. Murphy himself was out on the run with an assistant drafting lambs. Thrown to the ground, they held tightly to a fence to avoid being rolled about. They could see Napier across the sparkling blue bay. Smoke rising at Ahuriri 'brought home to them the real seriousness of the shake'.[81]

There was chaos on the Mohaka railway viaduct construction site where a caisson collapsed. 'Lofty' Young died under the rubble and a co-worker was buried to his waist. Other men on site took refuge in the foundation cylinders as the ground heaved. Not far away, C.A. Lawn was 'working with a survey party on the top of a razorback ridge when the whole hill moved' and the two sides slipped away from them. One man fell down the new slope but was unhurt. Another slip almost engulfed a car driving down the Wairoa road near Mohaka. Cliffs fell into the river, partially blocking the stream — it was later estimated that up to five acres (2 ha) of land collapsed into the river gorges. Elsewhere a Cook County overseer was standing near terraces just above the river when they erupted in a flurry of mud that 'covered the ground for six or eight feet, flowing eventually into the river'.[82] Another massive landslip blocked Te Hoe River, a tributary of the Mohaka, and a new lake developed behind the slip at Ngatapa, drowning farmland and several station buildings. The dam burst in 1938, releasing a rush of water that wreaked havoc further downstream.

In Wairoa, 50 miles (80 km) northeast of the epicentre, a span collapsed on the town's main bridge and shop frontages crashed down along the adjacent Marine Parade. Two people died — Mrs O'Malley, crushed by a falling tank, and Lim Kee Junior, hit by a falling beam as he was about to reach safety. The heaving earth split one house, toppled gravestones at the cemetery and ejected a corpse from the ground. Almost all chimneys were dropped, along with brick and unreinforced-concrete buildings. The post office tower fell with a clang of bells and rumble of disintegrating masonry. It was closely followed by the Gaiety Theatre; luckily only the projection staff were inside, and survived

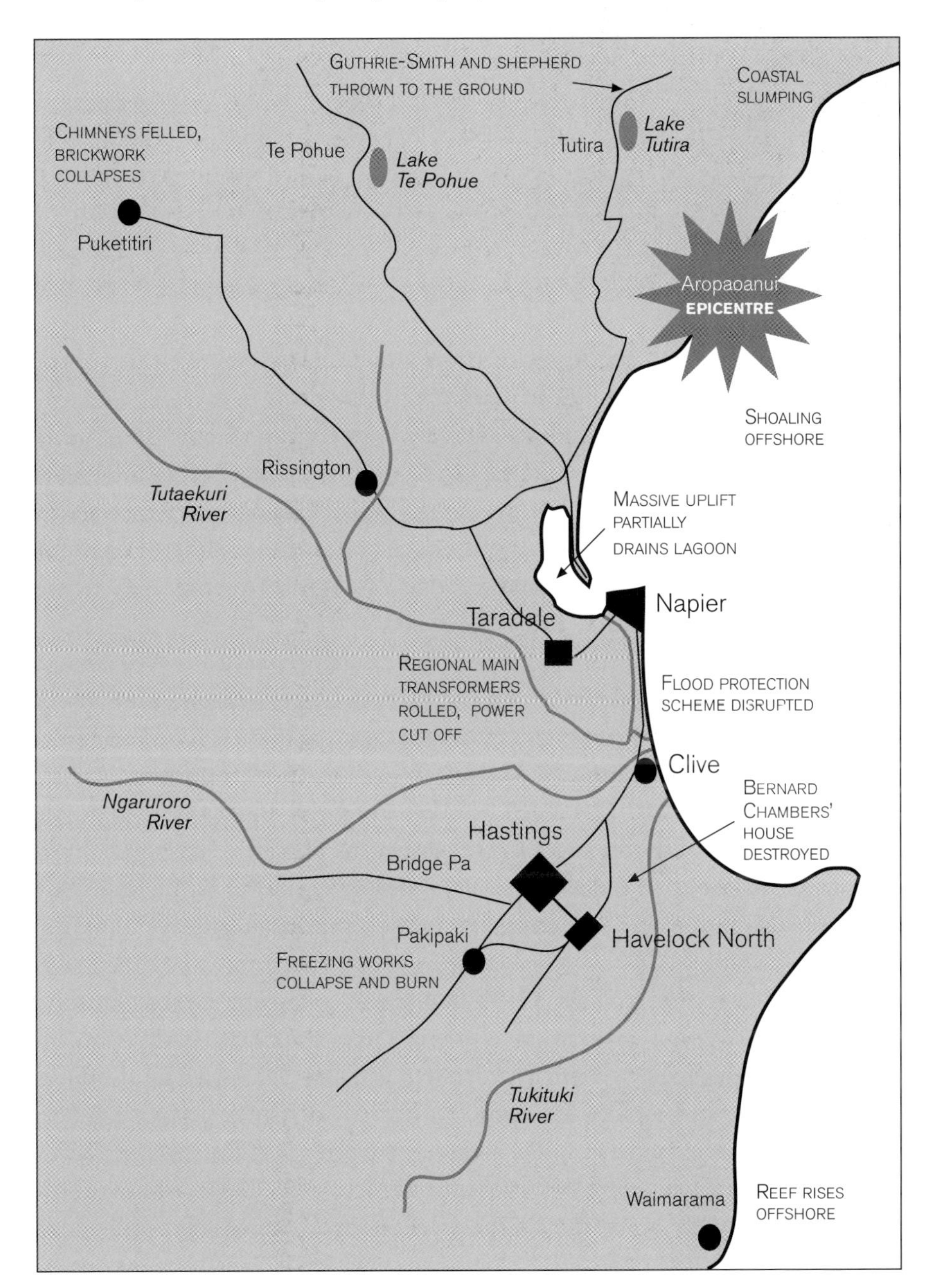

Some district effects, 3 February 1931.

The damage in Wairoa was almost as extensive as in Hastings. 'The Post Office tower was an awe-inspiring sight as it crashed to the ground in the middle of the shake,' one observer later wrote, 'crashing through the building with a terrifying sound of jangling bells and masonry. To those who visited the disaster it appeared almost certain that many of the officials had been killed, but miraculously they all escaped.'
(KITTY WOOD COLLECTION, PACOLL-1009-08, ALEXANDER TURNBULL LIBRARY)

without injury. In another stroke of good fortune, the children in the district school were being drilled outside when the earthquake struck.

Seven water tanks tipped their contents into the Wairoa Hotel, flooding the lounge, office and bar. A monstrous cocktail sloshed on the floor of the Wairoa Wine and Spirit Company, where 'dozens of cases of bottled beer, wines, spirits, cordials etc., were hurled from one end of the building to another'. Hogsheads of beer, each weighing a full 6 cwt (around 300 kg), were catapulted from a nearby gantry 'as if by giant hands'.[83]

A newly completed extension of the Clyde Hotel across the river collapsed, causing damage later estimated at £1500 (equivalent to about $108,000 in late-twentieth-century money, taking inflation into account). Type was hurled to the ground in the offices of the *Wairoa Star* and machines overturned. Luckily nobody was doused with hot metal from the linotypes, and staff began tidying up with the aim of producing a paper as soon as possible. Fire broke out in houses belonging to F. Hill and J. Millar, destroying both. As in Napier the local substation transformers jumped from their chocks, making restoration of power a slow task.

In Gisborne the roof of Hall's Buildings collapsed, almost crushing watchmaker R.T. Seymour and three others. Other buildings shed pediments, chimneys and decorations. Butcher T. Hibbert was almost crushed by a chimney falling through the roof of his shop.[84] The roof of Collet's Motors fell in, demolishing at least one car in the showroom below. Motorists in the Waikari Gorge west of the town were bombarded with rocks. R.H. Biggar, whose car had just crossed a bridge at the foot of the gorge, stopped the vehicle and had just got out when he was hit by a rock. The gorge collapsed around him, leaving the car standing on a piece of road like an island. He and his passengers climbed the hills in the hope of finding safety. Biggar later visited the Matahura Gorge, where he found the road had been obliterated.[85]

West of Napier, Puketapu farmer Ernest St Clair Haydon — whose property backed onto the Ahuriri lagoon — had gone to a neighbour's yard to help with sheep dipping. They were ready to start when the earthquake hit.

> Everyone held on ... as hard as he could and a boy who was sitting on a horse outside was thrown off his horse and the horse thrown to the ground ... the horse ... was trying to get on its legs the same as a drunken man might do ... the hills all around were absolutely opening and shutting and as the quake began to cease, great clouds of earth rose up — I should say to some 20 feet in height which looked as if all the hills had broken out into volcanoes ...[86]

There was little to destroy in many small inland Hawke's Bay localities such as Puketitiri. Wood-framed buildings such as these tended to stand up better than brick to the twisting forces of the quake; but in Puketitiri at least, what could collapse generally did. Puketitiri's origins as a milling village are evident in this pre-quake photograph of the late 1920s.
(DOROTHY SKINNER COLLECTION)

At Puketitiri, a tiny milling town 30-odd miles (50 km) west of the epicentre, a chimney came down in the Hendley mill, and the boiler shook loose from its foundations. At the Robert Holt & Sons mill, brickwork surrounding a newly installed boiler collapsed and a governor controlling the steam engine broke. It began to race and the engineers hurried to shut it down.[87] Curiously, the shock was scarcely felt in Kuripapango, at the southern end of the Kawekas; but at nearby Kereru, Hawke's Bay County Council chairman and farmer Frank Logan had just collected a mob of sheep when he found himself flat on the ground, watching the mountains dance.

Thomas Borthwick & Sons freezing works at Pakipaki was seriously damaged and then caught fire. Ten miles (16 km) further south, the brick-built parts of Te Aute suffered terrible damage, though the 'old wooden college church stood unscathed, with its spire erect, looking down on the devastation'.[88] In the small central Hawke's Bay towns of Waipawa and Waipukurau, wooden structures

Thomas Borthwick & Sons freezing works at Pakipaki, south of Hastings, was extensively damaged by the quake. Fire tore through much of the remains. Surviving heavy machinery was subsequently sent to Masterton.
(J.H. DAROUX COLLECTION, PACOLL-1009-01, PA1-F-145-77-73, ALEXANDER TURNBULL LIBRARY)

swayed and creaked as the shock rolled by. Brick and concrete shuddered, dropping decorations and pediments. Waipawa dentist G.E.T. Woods and his assistant hauled a patient to safety as the building collapsed. At Waipukurau, the post office tower shed debris into the street, a wall of the Tavistock Hotel fell out, and a number of shops around town were damaged.

Further south again the bucking ground twisted railway lines, while in Dannevirke — 80 miles (130 km) from the epicentre — chimneys crumbled and buildings cracked or lost their pediments. In Woodville, ten miles (16 km) further south again, residents heard a loud rumble before the ground heaved with a violence that toppled chimneys, broke glass and set telegraph poles and trees swaying.[89]

The quake affected the whole North Island. Lights swayed violently in Hamilton and the town clock stopped in Ngaruawahia. The shock drove people into the streets in Tauranga, swayed buildings at Te Puke, bent the steeple of St Stephens Church in Opotiki and broke water pipes in the town. In Wanganui a chimney toppled through the roof of the Metropolitan Hotel, shop windows were broken, and another falling chimney almost squashed Mrs Stiver of the

Jellicoe Block at Te Aute College after the quake. Like the ward in Napier Hospital the block was named after the First World War naval hero and post-war Governor-General of New Zealand, Admiral Sir John Jellicoe.
(J.H. DAROUX COLLECTION, PA1-F-145-70-32, ALEXANDER TURNBULL LIBRARY)

Waipawa was severely shaken by the earthquake and several buildings were heavily damaged. This is Bibby's Building.
(J.H. Daroux Collection, PA1-f-145-73-52, Alexander Turnbull Library)

Dannevirke, some distance southwest of Waipawa, was markedly less affected by the quake, but its brick buildings still suffered. This is Thomas Bain's premises.
(H.N. Whitehead Collection, PAColl-3068, Alexander Turnbull Library G-49459-1/2)

Kosey Tearooms. Glassware and crockery was broken at Tokaanu. People ran from their houses in Taumarunui, while at Taihape 'dozens of chimneys fell' and the railway bridge over Sulphur Stream near Tangiwai was found to be out of alignment. About 20 chimneys fell at Ohakune. The shock rolled south over the Wairarapa, stopping the Eketahuna town clock. In Carterton the quake was felt as a 'curious rotary movement'. [90] In Wellington, 200 miles (322 km) from the epicentre, people looked up as they felt the shock. Dogs howled and at least one chimney fell.[91]

The shock waves also swept over the South Island. H.F. Baird was 'measuring some records at a table in the Christchurch Magnetic Observatory Office' when he 'felt a slight movement which reminded one very strongly of the impression created by several aftershocks experienced at Arthur's Pass'.[92] At Westport there was a 'swaying motion' that lasted more than a minute, and even in Invercargill the quake had enough energy to swing electric lights on their cables.[93] The shock waves swept around and through the earth, ringing the planet like a bell and triggering seismographs in — among other places — Bombay and Calcutta, as well as at Kew Observatory in London.[94]

CHAPTER 2

'Sand and water is not very sticky'[1]

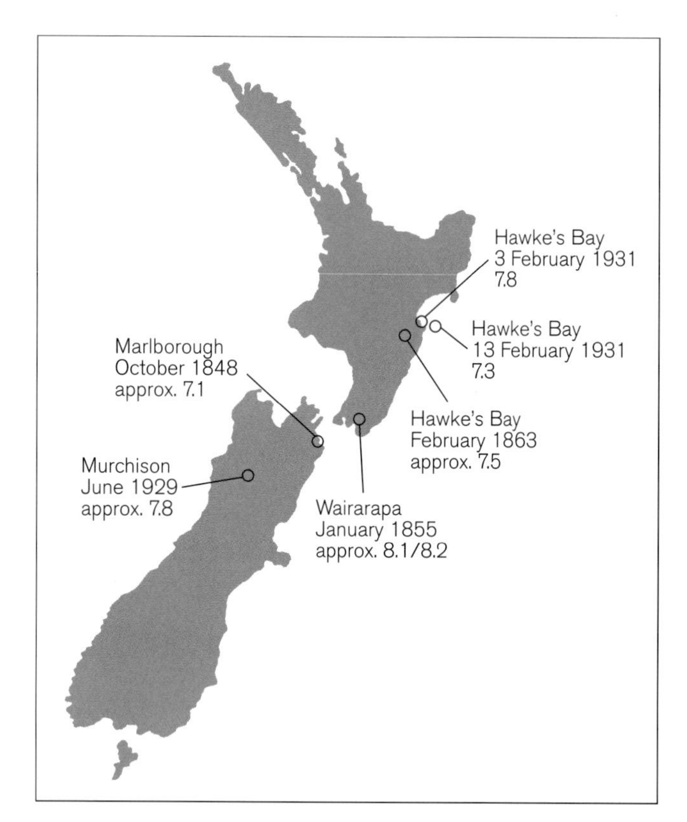

Some of New Zealand's larger earthquakes, 1848–1931.

Earthquakes had been part of life in Hawke's Bay since the earliest days of European settlement. Many of the settlers who arrived during the 1850s had already lived through earthquakes in Wellington — Petone settlers felt their first quake as early as May 1840. Major shocks there in October 1848 delayed Fred Tiffen's first attempt to run sheep to Hawke's Bay.[2] There was extensive damage to chimneys and brick buildings.

> Only 1 bakers [sic] oven was left intact ... A brick wall fell and killed Sgt. Lovell and 2 children. Medical hall kept by Dr Dorset became a scene to be imagined with bare shelves and the contents broken and badly mixed ... A number of land slips occurred on the wooded hills between Wellington and Wairarapa and in one instance a house was shaken off the piles supporting it.[3]

Some settlers blamed poor mortar. 'Sand and water is not very sticky,' visitor Charlotte Godley explained in a letter to her mother.[4] The quake was centred on the Wairau Valley, and later estimated to have a magnitude of 7.1, with a strength in Wellington of about VIII on the Modified Mercalli Scale of felt intensity.[5] Wellington swayed to another tremor in May 1850. The proverbial 'big one' hit in late January 1855, a major failure of the Wairarapa fault with an estimated magnitude of 8.1 or 8.2, and a peak intensity in Wellington of X on the Modified Mercalli Scale.[6] Destruction spread from Wellington to Wanganui, and the quake was felt as far north as Wairoa.[7]

These experiences offered lessons about unreinforced masonry, but quake-proofing did not feature much in colonial construction. Complacency and lack of experience were part of the issue, but another reason seems to have been that the flurry of quakes during the 1845–55 period tailed off. Nobody understood the real mechanisms behind seismic activity, and later comments indicate that many settlers believed the worst was over. Fred Tiffen's description of the 1848 Wellington earthquake as being 'not such as people make a song about now a days but real chimney levelling earth-cracking inspiring articles' is a case in point.[8] Another cluster of quakes began in the 1920s, but the fact that this was a new pattern was not obvious.[9]

HAWKE'S BAY EARTHQUAKES 1850–1931

Pioneer missionary William Colenso — a man of towering intellect and pugilistic temperament — left the earliest known notes of a Hawke's Bay quake in 1850.[10] Town land sales in Napier took place three months after the Wellington 'big one' of 1855, and by the end of the decade the provincial town was a promising community, though underdeveloped because the huge undermanned pastoral runs of the hinterland did not do much for urban growth.[11]

Provincial authorities wrestled with strategies to balance the economy,[12] but skewed development meant there was little to smash when a massive earthquake hammered the region 'a few minutes past one' on the morning of 22 February 1863. It was centred southwest of Napier, with a magnitude of at least 7.5, and was literally a wake-up call. From Napier to Waipukurau, settlers bleary with sleep huddled on floors as their houses twisted, sending pots and pans skittering across kitchens, books flying from shelves, and furniture on a mad dance across floors. The *Hawke's Bay Herald* reported that in 'hotels and stores':

> ... the contents of whole shelves [have] ... been swept to the ground and destroyed ... Innumerable chimneys have broken off at the roof, and, in one or two instances, the bricks came through ... The barrack chimneys have all gone, also one side of the mud wall which surrounded the reserve. It was, however, a feeble structure at the best of times. The road to the Spit exhibits several cracks, but it having been all new ground at a comparatively recent period renders this an unimportant fact ... The atmosphere, we may add, was not affected — Mr. Koch having had presence of mind to examine his instruments at the time.[13]

However, although people 'felt rather queer during its continuance', the quake itself was, in the opinion of the paper, 'quite a minor affair' by comparison with the Wellington shock of a few years earlier. The news took third place after a report about the withdrawal of Imperial troops and an account of the voting to replace Colenso on the Provincial Council.[14]

The quake may not have been exceptionally strong in Napier, but elsewhere appears to have peaked at IX or X on the Modified Mercalli Scale, making it similar to the 1931 event. Walter Slater, who lived through both, certainly thought it was comparable, and William Nelson came to the same conclusion after comparing the scars both left on the landscape.[15] At Wakarara, pastoralist Hector Smith was woken by:

> ... the terrific rattling of doors, windows etc. I was nearly thrown out of my bed. I soon sprang out however & rushed outside, but on the ground I could not stand I had to sit down! it shook so — indeed I thought the time of the end had come — the shake continued so long. The birds fell from their perches and the fowls made a most unearthly noise. All chimneys, as a matter of course, were thrown down — the peculiar sensation I felt at that time ... I can hardly describe, I felt as if sitting on boiling water — the ground seemed to bubble so, beneath me. The spurs & creeksides were rent so as to preclude & slur the usual tracks & the soil on said spurs seemed as if ploughed and harrowed. And one settler had the Nails of his house (those fastening the weatherboards) draw about 1 inch out of the studs. The mud in the different swamps was shaken up so that the formerly clear creeks became quite discoloured.[16]

When this picture was taken around 1865, Napier was the provincial capital and could have been transplanted from the American west of the same period. Hastings Street, nicknamed the White Road, dominates; the large building where it kinks is the original Masonic Hotel. The Bank of New Zealand is just visible on the right-hand side of Hastings Street near the hill, and swamps extend into the distance beyond the triangle of flat land. A major earthquake had rattled the town just a few years before this picture was taken, but the wooden structures essentially shrugged off the effects, and in the absence of a workable theory to explain quakes, settlers had no way of knowing there would be more.
(*Daily Telegraph* Collection)

Archdeacon Samuel Williams subsequently argued against the Napier cathedral being built in brick as a result of the experience, but — as J.G. Wilson dryly put it in 1939 — 'his warning was not heeded'.[17] The decision had tragic consequences but, to most people, precautions seemed unnecessary. 'Roman Cement', the pseudonymous 'A. Brick' advocated in March 1863, would be enough to build quake-proof chimneys.[18] In any case, later earthquakes had nothing like the intensity of the 1863 event. A tremor in March 1890 stopped clocks, and a strong jolt was recorded in Napier in August 1894, but neither did much damage. Architects in Napier and Hastings designed new buildings in masonry or concrete with barely secured decorations. These were tested to the limit during the morning of 9 August 1904 when Napier residents felt:

Napier's Anglican Cathedral was built in brick against the advice of Archdeacon Samuel Williams. The magnificent structure collapsed completely during the 1931 earthquake, killing or injuring those attending morning service.
(STORKEY COLLECTION)

> ... a tremor, followed by a roaring sound as of distant artillery. At the same moment came a tremendous shock that made several buildings totter, most to their foundations. The fronts of apparently substantial structures swayed to and fro, massive telegraph poles in the main streets waved like willows in a strong wind and flagpoles quivered as storm tossed reeds.[19]

Napier's Shakespeare Road, the main route up the Napier hill, around 1900. Napier had lost its shanty-town appearance but stylistically still harked across the Pacific to the United States rather than Britain. Many of the wooden buildings here were still in use when the 1931 quake struck.
(*DAILY TELEGRAPH* COLLECTION)

This was another major shake, centred near Cape Turnagain, which caused serious damage to inland Hawke's Bay towns and felled chimneys in Wellington. The strength in Napier has been estimated at VI or VII on the Modified Mercalli Scale.[20] Napier was rocked again in June 1921 by a tremor that peaked at between VI and VII on the Modified Mercalli Scale.

Despite these lessons, architects continued to add embellishments to new public buildings, and it was the 1920s before ferroconcrete began to feature as structural material. Private homes were put up without much thought for quakes, and while most were wood-framed, which could twist and sway in all but the strongest quakes without collapsing, others were brick or concrete and posed considerable risks — especially on the newly reclaimed suburban sections of Napier South, founded on barely compacted silt. Part of the problem was that nobody understood how earthquakes worked — it was the 1960s before the underlying tectonic processes were properly explained.

One of the few dissenters was Havelock North pastoralist Mason Chambers. When his home burned down in 1914 he insisted that the replacement had to be fully quake-proofed, despite the difficulties of importing reinforcing steel in wartime. Auckland architect W.H. Gummer came up with a cubist masterpiece, dubbed Tauroa, which withstood the 1931 quake with ease. Mason's younger brother Bernard Chambers was not so lucky. His 1922 Spanish Mission mansion, designed by talented local architect William Rush, collapsed completely in 1931.[21]

Tauroa, home of leading Havelock North pastoralist T. Mason Chambers, survived the 1931 earthquake with virtually no damage. There is little to distinguish this post-quake picture by James Daroux from pre-quake images. Tauroa was one of the most advanced and forward-looking homes in New Zealand when it was built during the First World War. Auckland architect W.H. Gummer sketched out several designs in 1914 and a final plan was approved by Chambers the following year. Construction was complicated by Chambers' insistence on quake-proofing – the problem was getting reinforcing steel during wartime. 'I could import this quite easily as I have a London agent,' Chambers told Gummer, 'if it were not for the delay of about three months that would ensue.' The house was finished in 1916.
(J.H. DAROUX COLLECTION, PA-F-145-69-27, ALEXANDER TURNBULL LIBRARY)

DECADE OF STYLE — THE SWINGING TWENTIES

The Hawke's Bay that crashed to ruin in 1931 was very different from the Hawke's Bay of even 20 years earlier. The quake is often viewed as the catalyst for change from 'colonial' to 'modern' city centres in both Napier and Hastings. In fact, the physical and social transformation of Hawke's Bay began years earlier, and provides a context by which we can understand the social response to the quake.

Modernism in its broadest sense was an eclectic range of artistic, architectural, musical and literary styles that flourished during the first decades of the twentieth century.[22] Its enduring legacy was its architecture, which exploited the new techniques and materials of the second industrial revolution, and was characterised by a wide variety of bold forms, including classical revival, Chicago school, Bauhaus, cubist, futurist, constructivist, expressionist, streamline modern and ultimately the hybrid 'moderne'. Some of these disparate styles were later grouped under the generic title 'art deco'.[23]

Spanish Mission — a product of 1890s California and one of the antecedents of modernist architecture — was in vogue around Hawke's Bay by the First World War. One of the first large Spanish Mission buildings in the area was the Hastings Municipal Theatre, designed by Henry White in 1914–15.[24] Albert Garnet was another prolific exponent of the style; his work included Kilford and Ebbett's building of 1915, the Muncipal Buildings of 1917, and the Fitzpatrick & Co. building of 1924. In 1913 William Rush designed Iona College in Spanish Mission style, following this with Bernard Chambers' new home, completed in 1922.

Children enjoy a summer dip in the Tutaekuri River, then flowing past the bottom of Napier's Ellison Street and slated to stay that way despite fractious efforts to divert it into the sea south of the town. The partial diversion in vogue by 1931 was completely disrupted when the earthquake lifted the land.
(DAILY TELEGRAPH COLLECTION)

This picture graphically shows how Napier South was developed around the turn of the twentieth century. The Tutaekuri River flows directly past the sports grounds, diverted to provide silt that could then be spread over the swamp beyond. The first sections were on sale before 1915, but the area was barely compacted and particularly vulnerable to earthquake.
(DAILY TELEGRAPH COLLECTION)

Spanish Mission was not the only new style in the district by this time. Havelock North architect James Chapman-Taylor championed the pastoral-modern look of the English 'arts and crafts' movement during the same period. For philosophical reasons he combined attractive exteriors with monolithic reinforced-concrete construction — his houses had walls a foot (30 cm) thick and in the minds of some were not dissimilar to bunkers. When C.T. Natusch inspected the eleven Chapman-Taylor houses around the district after the quake in 1931 he found only minimal damage. W.H. Gummer, meanwhile, explored American styles. He followed Tauroa with a magnificent home for the van Asch family at Craggy Range in 1919, and a classically influenced structure for Maurice Chambers in 1926.[25]

Modernism came to Napier at the end of the First World War. Frank Lloyd Wright provided inspiration for J. Louis Hay's 'prairie' style Soldiers Club, completed in 1920. Hay adopted 'prairie' styling again for the Women's Rest in Clive Square, and his nearby Central Fire Station echoed Wright's Larkin building of 1903.[26] Eric Phillips adopted conservative 'classical revival' styling for the Public Trust Office on the corner of Dalton and Tennyson Streets during the early part of the decade. It was completed in 1926 and was one of the earliest large reinforced-concrete buildings in the district — Phillips apparently specified railway line to bolster the Doric columns.[27]

Renowned for its acoustics and the size of its stage, Napier's Municipal Theatre opened in 1912.
(STORKEY COLLECTION)

Twentieth-century styles were in vogue in both Napier and Hastings for years before the 1931 earthquake struck, setting the stage for the proliferation of various modernist stylings that featured in post-quake reconstruction. One of the notable survivors from the immediate pre-quake era was Napier's Public Trust Office, seen here in early 2000. This classical revival masterpiece also escaped the fire, and during the days after the quake stood like a lone beacon of order amid the ruins. It did not escape entirely undamaged, however; its original brick panels were replaced in concrete.
(MATTHEW WRIGHT)

Many of the foreshore entertainments for which Napier later became known were not developed until after the quake. Uplift extended the beach and made shore-front buildings easier to construct without fear of inundation. Until then, waves tended to slosh into the town from time to time. The courthouse, visible on the right-hand side of the Marine Parade diagonally opposite the baths, was damaged in the 1880s when a monster wave mounted the sea wall. (STORKEY COLLECTION)

This pre-quake view of Napier's Memorial Square, with the cenotaph, fire station and hill in the background, was one of more than two dozen photographs purchased by Napier bookseller W.E. Storkey in 1931 and published as souvenirs. (STORKEY COLLECTION)

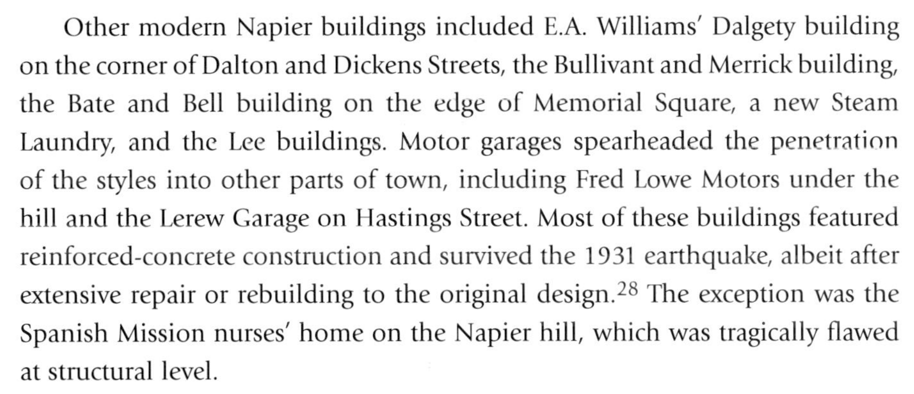

Other modern Napier buildings included E.A. Williams' Dalgety building on the corner of Dalton and Dickens Streets, the Bullivant and Merrick building, the Bate and Bell building on the edge of Memorial Square, a new Steam Laundry, and the Lee buildings. Motor garages spearheaded the penetration of the styles into other parts of town, including Fred Lowe Motors under the hill and the Lerew Garage on Hastings Street. Most of these buildings featured reinforced-concrete construction and survived the 1931 earthquake, albeit after extensive repair or rebuilding to the original design.[28] The exception was the Spanish Mission nurses' home on the Napier hill, which was tragically flawed at structural level.

Hastings was not far behind. The town had led Hawke's Bay's stylistic revolution with its Spanish Mission buildings, and architects lost no time exploring other new ideas after the war, notably expressed in Edmund Anscombe's Hawke's Bay Farmers building. Albert Garnett's stripped classical 'Villa d'Este' was completed on Heretaunga Street in 1929. The trend looked set to continue into the 1930s in both towns. Government architect J.T. Mair designed a classically influenced block for Napier Girls High School in 1930. By the end of that year plans were afoot for another large modernist structure, the Market Reserve building, on the corner of Hastings and Tennyson Streets.

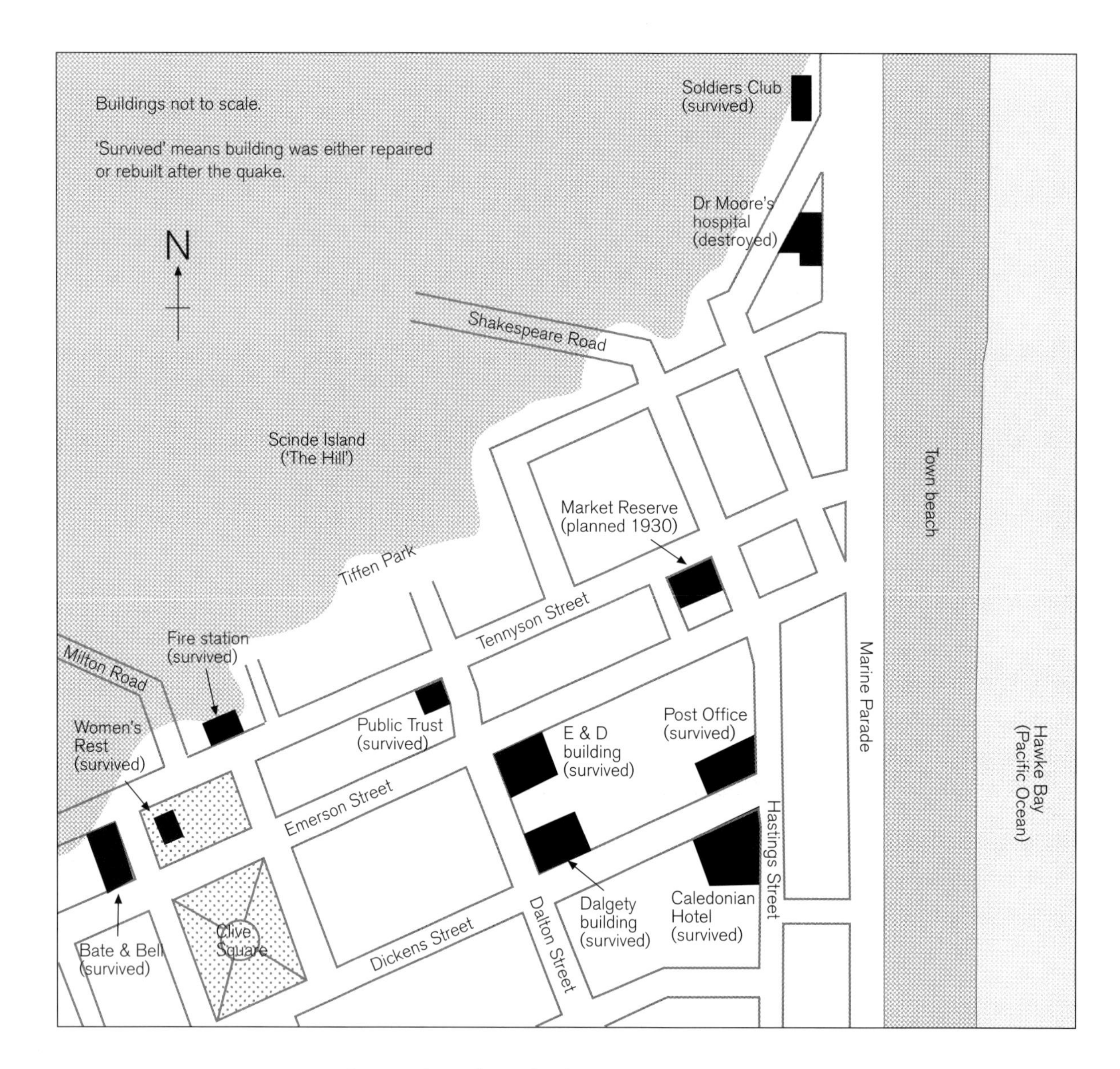

Some pre-quake 1920s and modernist buildings in Napier.

These styles reflected a changing society. New Zealand had never experienced an 'Edwardian' period as such. Instead the colony, with its near-caricature of extreme Victorian economic ideals and its huge collective inferiority complex, was dragged abruptly into the modern world during the First World War. A large proportion of New Zealand's youth served overseas, developing a unique camaraderie of service life while being exposed to overseas culture *en masse* for the first time. Returning soldiers were fired with ideas and felt a sense of national unity that their colonial fathers had never known. The result was a heady social brew that was instrumental in shaping post-war New Zealand.

This view looking west from the corner of Railway Road and Heretaunga Street dates to about 1920 and captures something of the flavour of pre-quake Hastings. The late settler-era shops and pillared verandahs are noteworthy.
(S.C. Smith Collection, PAColl-3082, Alexander Turnbull Library G-48216-1/2)

Havelock North was nicknamed 'the village' by its residents, a close-knit community where art, literature and independent thought flourished during the first decades of the twentieth century. The small 'arts and crafts' building partly obscuring the Whyte and Glenny premises is the Transformer House and Clock Tower, designed by avant garde architect James Chapman-Taylor in 1914. Its bunker-like construction survived the quake with ease and became a well-known village landmark.
(S.C. Smith Collection, Alexander Turnbull Library G-48876-1/2)

Albert Garnett's stripped classical 'Villa d'Este' was finished less than two years before the 1931 earthquake tore through central Hastings. The original structure was badly damaged during the quake and had to be demolished. A replacement went up afterwards to the same design, while similar 'new for old' structures were also built in Napier, all contributing to the post-quake look of both towns.
(Matthew Wright)

This society was particularly receptive to the new styles flourishing worldwide during the 1920s. New technology — radio, gramophones and movies — combined with popular picture magazines to bring modernism to the mainstream. Clothing and hairstyles were transformed. Music hall standards went out of the window in favour of swing, and while 1920s jazz arrangements had yet to gain the lushness of later decades, the relaxed rhythms marked a sharp break from on-beat music hall plunkiness.

Cinemas proliferated, buoyed by studio marketing strategies that made the actors as much a focus of attention as their work. In Hawke's Bay even small towns such as Takapau, Ormondville, Waipawa and Waipukurau had their own movie theatres by the 1920s. Napier had four. A regular Hollywood gossip column in the *Daily Telegraph* offered glimpses into the lives of Douglas Fairbanks, Mary Pickford, Charlie Chaplin and other stars. Readers were tantalised by accounts of stylish social evenings where sharply dressed revellers were chauffeured to endless parties, dinners and dances.

Messing about in boats was a popular pastime in the lagoon northwest of Napier. The lagoon was also a traditional resource for Maori, but all this came to an abrupt end with the earthquake, which drained much of the water.
(DAILY TELEGRAPH COLLECTION)

These fantasies offered a welcome diversion from the realities of 1920s New Zealand. Times were hard: the decade only 'roared' for the United States. Britain had been bled economically dry by the war, and New Zealand, which relied heavily on 'the old country', was in dire straits. A brief post-war upturn dissipated during 1921–22, and apart from a slight surge in 1926–28 the decade was one of gloom for most New Zealanders as the economy lurched from downturn to downturn. The 1929 stock market crash was the last straw: New Zealand collapsed into an economic free-fall that continued until the bleak winter of 1933.[29]

In early 1931 New Zealand was in the middle of this steady decline, which continued unabated despite the efforts of a new coalition government to halt it. Indeed, it has been argued that government initiatives of the day — a puritanical throw-back to the 'classical' economic dogma that had already failed during the colonial era — actually intensified the downturn.[30] A summer drought at the end of 1930 compounded the crisis for Hawke's Bay, undermining the horticultural industry on which the region relied. The earthquake came at the worst possible time.

Chapter 3

'The glare ... lit up the sky'[1]

Each man in those days had his own particular impossibility to relate. 'And miracles don't happen,' somewhere says Mathew Arnold, clinching his argument. They do in earthquakes.

H. Guthrie-Smith[2]

As the quaking subsided on that fateful February day in 1931, people from Waipukurau to Wairoa — but mainly in Napier and Hastings — lay trapped and pinned by fallen debris, crushed, bruised and in many cases critically injured. Others had been hit by flying bricks, glass, wood, furniture or other objects.

Among the first to react were old soldiers, servicemen who had fought on the western front a decade and a half earlier. For them the wreckage of Napier and Hastings was an all too familiar sight. They knew what to do. Former British soldier F.C. Wright was working on the hill when the quake struck. Reaching town, he realised the enormity of the disaster and at once joined the rescuers. He did not reach home until after 6.00 p.m. — by which time his worried family were sure he had been killed. He later compared the destruction in Napier to what he had seen in French villages bombarded by shellfire.[3]

Others also responded quickly. Engineers in the Napier Gas Company plant shut off the main distribution valves within three minutes, averting the risk of gas-fuelled fires in the shattered town. P.W. Barlow left his devastated Napier office immediately after the earthquake to find his staff outside 'standing in a semi-dazed condition'. He organised them into groups of six to lend help wherever they could.[4]

Fire and rescue — Napier

As the dust settled in Napier, commercial photographer A.B. Hurst hurried outside to record images of the shattered buildings. Yet for Napier the destruction had only just begun. Chemist R.S. Munro, managing the Friendly Society dispensary at the lower end of Emerson Street, returned to his shop minutes after the quake to find the twisted building on fire. The brigade, with its four permanent and 25 volunteer firemen, was not far away — the new fire station stood just one block distant, its alarm bells clanging after a short circuit had set them off. The station broadly withstood the shake, but the earthquake rolled the heavy No. 1 Dennis engine through the wooden doors, almost crushing E.E. Symons in the process and jamming the starting handle in a tangle of broken wood. It was some minutes before the men could get it going.

When Fire Superintendent W.J. Gilberd and Station Officer J. Driberg arrived at the Friendly Society dispensary they could only get a 'trickle of water'[5] and, driven back by fumes, were unable to save the building. Then reports came in that W.R. Henderson's Hastings Street pharmacy had ignited. Gilberd left two firefighters working on the Friendly Society fire and took the No. 2 Dennis to Hastings Street. The blaze there had actually started in the rear of Henderson's shop midway between Browning and Herschel Streets, adjacent to the County Council offices.[6] Here there was a 'good working supply and pressure of water', and the brigade was 'getting Henderson's fire well in hand' when Arthur Hobson's shop in the Masonic Hotel block also erupted into flame,[7] joining a fire that had apparently started in the ruined kitchen of the hotel.[8]

Hobson's premises sat under part of the hotel — bedrooms were built over the top of his pharmacy — and with the ruined hotel already ablaze there was every chance the whole block might go up. Just then the water pressure died, forcing the firefighters to use the pumps on the engine. The No. 1 Dennis arrived after dealing with the Lower Emerson Street fire, which had been contained to the Friendly Society building. Both engines were then used to fight the Hastings Street flames. With a prevailing wind blowing the fire out to sea there was hope the fire could be contained in the Hastings Street–Marine Parade blocks. However, a shift of wind sent the flames roaring up Hastings Street towards the hill. The water ran out altogether, and in the face of the advancing inferno the brigade had to 'abandon their position, losing a quantity of hose and gear' but saving both engines from immolation.[9]

Rescuers hastened to save trapped victims from the fire. Sixty-one-year-old Edith Mary Barry was one of nine faithful attending the morning communion service led by Archdeacon Brocklehurst in the Anglican Cathedral when the earthquake struck and the building collapsed. Her son knew she was somewhere under the rubble and found her almost completely entombed, her legs and body pinned by a heavy girder.[10] Willing rescuers tried to prise her free, but 'all efforts to lift the huge girder were unavailing'.[11] Flames from the Hastings Street fire spread to the ruined cathedral and 'gradually drew nearer and nearer to the stricken woman'. Firemen played hoses on the wreckage until the water ran out. The situation was desperate. Dr G.E. Waterworth was on the scene, but all he could do to save the agonised woman from being burned alive

Couples stand arm-in-arm as the fire brigade tackles the blaze in Henderson's pharmacy. Jets rising two and three storeys provide graphic evidence of excellent initial water pressure. Debris from cracked and damaged buildings almost blocks the road – a further hindrance to firefighters. The Masonic Hotel is to the right of the photographer in these views down Hastings Street from the Emerson Street corner, looking towards the hill. A.B. Hurst captured these images perhaps 20–30 minutes after the quake, and the Masonic block fires that did all the damage appear not to have started at this stage.
(Storkey Collection)

The same scene from the other direction, taken a few minutes later from the intersection of Browning and Hastings streets. Water pressure is good for the moment and the brigade appears to have the Henderson's Pharmacy fire well in hand.
(K.S. Williams Collection, PA Coll-1960-11, Alexander Turnbull Library)

The fire that tore through the central business district started in the Masonic block, probably in two places – Arthur Hobson's pharmacy and the Masonic Hotel kitchen. Firefighters were in dire straits after the water ran out. In this classic image of the disaster, a fireman – apparently Superintendent Gilberd himself – prudently withdraws through the debris towards Shakespeare Road, pursued by the fires from the Masonic block, fanned before a brisk breeze. Although it is often captioned 'ten minutes after the quake', this photograph was in fact taken about three-quarters of an hour later. The wind pushed the blaze up Hastings Street towards the site of the earlier fire in Henderson's pharmacy, which is behind the photographer to the left. Where the couples stood earlier has been engulfed by a wall of smoke and flame.
(P.T.W. Ashcroft Collection, Alexander Turnbull Library F-139885-1/2)

ABOVE: The fire has advanced half a block and reached the intersection with Browning Street, on the edge of the hill housing district. The Bank of New Zealand burns fiercely behind the power poles. Firemen fought on here until the last moment in the vain hope of halting the blaze. This fire sealed the fate of Edith Barry, trapped in the ruins of the church adjacent to the Bank of New Zealand, down Browning Street to the right.
(KITTY WOOD COLLECTION, PACOLL-1009-02, ALEXANDER TURNBULL LIBRARY)

RIGHT TOP: The fire rips past in this view looking seawards towards Hastings Street, advancing from right to left of the picture towards Shakespeare Road and the hill.
(K.S. WILLIAMS COLLECTION, PA COLL-1960-12, ALEXANDER TURNBULL LIBRARY)

RIGHT BOTTOM: For a while it seemed that the fire would tear into Shakespeare Road and wreak havoc in the vulnerable wooden hillside housing district beyond – Brewster Street, Corry Avenue, Madiera Road and Shakespeare Terrace. Herculean efforts by fire-fighters, naval 'bluejackets' and residents, coupled with a timely change of wind, halted the fire before it could get too far up the hill. The colonial-era wooden buildings at the base of Shakespeare Road were scorched but did not catch alight.
(J.H. DAROUX COLLECTION, PA1-F-145-14-84, ALEXANDER TURNBULL LIBRARY)

was inject her with what reporters called 'a large dose' of morphine.[12] 'Death was inevitable,' the *New Zealand Herald* reported two days later, 'but at least, it was without pain.'[13]

One other person died in the wreck. The rest were dragged from the cathedral with fractures, bruising, contusions and multiple injuries, including Archdeacon Brocklehurst, with serious back injuries — there were early fears that 'his injuries may prove fatal'.[14] People caught in other collapsed buildings or hit by debris suffered various injuries. Arthur Spackman survived with a broken leg; Albert Lenden had a broken leg and head injuries; Bessie Crowley had broken legs and arms. With only a case of concussion, Bank of New South Wales manager W.J. Leversedge came off relatively lightly by comparison.[15]

From a vantage point above Cathedral Lane the tumbled and burned remains of the cathedral presented a dismal sight after the quake.
(STORKEY COLLECTION)

Doctors 'seemed to spring from nowhere',[16] rushing from their surgeries and private hospitals to help.[17] Rescuers quickly organised trucks to take the casualties to the hilltop hospital. However, the commandeered vehicles reached Napier Terrace to find another calamity unfolding. Horrified rescuers were swarming over the ruined nurses' home in the hope of extracting survivors, while a steady stream of doctors, nurses and orderlies were wheeling patients from the ruins of the hospital beyond. Parts of the only base hospital in the district were no more than wreckage — including the new Jellicoe Ward. Other wards were upright but clearly unsafe, including the surgical block, which could not be used until it had been checked.

Although evacuation had to proceed past the dusty ruins of the home where nurses lay dead or dying, hospital pharmacist J.S. Peel noted a 'complete absence of panic'.[18] Criton Smith was very admiring of the nurse who had run from the collapsing home less than an hour before. 'Although temporarily dazed by her experience, she quickly went to the aid of the other nurses brought out from the home and she has not been to bed yet,' Smith told reporters later in the day. 'That is typical of the way all those girls have worked.'[19]

Dr A.G. Clark organised an emergency surgical station in the Botanical Gardens. An operating table was put under an archway at the top of the gardens,[20] and within an hour life-saving operations were being conducted with full sterilisation and anaesthetic procedures. Aftershocks rocked the ground as the doctors worked. Clark had a nurse alert him while he worked:

when she called 'Stop' he lifted his hands and waited for the shock to pass. Initial news reports suggested that some emergency surgery had been done without anaesthetic,[21] but this was untrue and a few days later Hastings doctors reiterated to reporters that 'every operation, major or minor, had full surgical anaesthesia'.[22]

Casualties far exceeded the capacity of emergency facilities. 'All we could do was to lie them on the lawns to wait their turn for treatment,' Sister Mary Eames later wrote.[23] Off-duty doctors and nurses who had been in town quickly returned, among them one nurse who tried to help schoolchildren during the earthquake itself. Later she recalled that:

> ... by degrees surgical stores, drugs etc. were extricated from the ruins ... all Tuesday we worked like war nurses ... We were washing wounds and dressing them and pumping in injections.[24]

Trim, tidy and vital – Napier Hospital during the 1920s. Buildings from left to right are the Margaret Ward, Shrimpton Ward, administration building, Stokes Ward and Moseley Ward. The destruction of this, the only base hospital in the district, was a severe blow.
(ALEXANDER TURNBULL LIBRARY F-60984-1/2)

Local residents pitched in to help. George Brown arrived at his Napier hilltop home to find his wife, Jean, and two daughters safe. When doctors came looking for sterilised water he lit their 'camp cook' and 'kept kettles going for tea, of which large numbers of people gratefully partook'.[25]

A major rescue effort focused around the nurses' home. There was no hope for the three clerical staff on the lower floor, but the nurses 'were placed in a slightly better position and it was thought that some at least might be saved'. Two were found trapped by a fallen slab of wall and collapsed staircase. A dozen

Napier Terrace outside the hospital was soon jammed with patients, doctors, nurses and helpers as evacuation of the ruined buildings got under way.
(ALEXANDER TURNBULL LIBRARY F-29566-1/2)

rescuers spent three hours trying to free them. Every effort proved fruitless, and in the end the slab had to be broken with sledgehammers. Efforts to relieve the effects on the injured nurses by wedging the slab with crowbars while it was hammered were only marginally successful, but they 'stoically endured the long suspense, crushed in discomfort and pain'.[26] Six nurses were pulled from the debris, seriously injured but alive.

The well-tended foliage of the Botanical Gardens adjacent to the hospital provided an unlikely setting for patients. Fortunately the weather was fine and warm, a typically brilliant Hawke's Bay summer's day.
(ALEXANDER TURNBULL LIBRARY F-60975-1/2)

The remnants of the nurses' home on the corner of Chaucer Road and Napier Terrace. Fatally flawed at structural level, the magnificent Spanish Mission building fell in a manner likened by most observers to the collapse of a house of cards. After the quake a new wooden nurses' home was built slightly downhill from the hospital; a permanent replacement, diagonally opposite the hospital on Napier Terrace, had to wait until the 1950s.
(ALEXANDER TURNBULL LIBRARY F-57116-1/2)

The Jellicoe Ward, Napier Hospital, after the quake. The loss of Hawke's Bay's only major hospital demonstrated the folly of placing all the regional medical services at one location.
(Photographer unknown, Alexander Turnbull Library F-17204-1/4)

The remains of the hospital after the earthquake.
(PAColl-1960-07, Alexander Turnbull Library)

Not all casualties arrived at the hilltop hospital; nearly a hundred were taken to the Byron Street police station, where they were treated by Surgeon Lieutenant-Commander McVicker from the *Veronica*.

Plans for an emergency field hospital were quickly dusted off. The scheme had been devised several years earlier by Dr Allan Berry after a central Hawke's Bay train accident flooded Napier Hospital with casualties. Thinking of a civil disaster, he had proposed siting an emergency hospital at the Napier racecourse, because it had water and was far enough inland to be out of reach of a tsunami. Four surgical teams were on site by mid-afternoon on 3 February, though it was the next day before the hospital was fully set up. That did not stop emergency surgery. Doctors worked under the stark glare of car headlamps until 2.00 a.m. A dressing station was also established in McLean Park. A total of 454 wounded were tended in Napier and Hastings. Some 333 patients were subsequently evacuated to Wanganui and Palmerston North.

Sounding the Alarm

Five of the six hawsers tying HMS *Veronica* to the wharf snapped during the quake, and when water rushed from the inner harbour like a tidal race the ship hit the bottom, threatening to cant over if the last hawser parted. Commander Morgan rallied the crew to secure the ship. The valves in the *Veronica*'s receiver were smashed when the ship was hit by the harbour bottom, but this did not prevent transmission and at 10.54 a.m. Morgan signalled his commander-in-chief in Auckland by morse.

While this exchange was carried out Morgan made arrangements to land parties of men with medical supplies, food and rescue material.[27] The first two squads left under Lieutenant Warrand and Sub-Lieutenant Vesey, closely followed by a third under commissioned gunner-in-charge Gale. A first aid party under Surgeon Lieutenant-Commander McVicker left the ship soon afterwards, and a fifth group marched round the hill to the Marine Parade foreshore where they set up a food depot. At about 1.30 p.m., two-and-a-half hours after the disaster, Morgan sent his last remaining officer, Lieutenant-Commander Grimes, to liaise with local authorities. During the afternoon these parties were joined by merchant seamen from the *Taranaki* and *Northumberland*, who arrived to put themselves under Navy authority.

Morgan and Signalman Randall meanwhile went ashore to find parts for the receiver. They checked L.F. Peach's general store, opposite the narrow dockway known as the Iron Pot. Peach had a wireless, made by Tom Frater for his son Frank. The contents of the house had been jumbled — even the piano lay on its side — but the wireless was intact, along with the fragile valves the Navy needed.[28] It would appear that the receiver was back in action by early afternoon, allowing Morgan to respond to worried queries from Cabinet and to update the government on what was happening.

One of the enduring myths of the quake is that the *Veronica* raised the first general alarm. In fact, while the *Veronica* played a key role in alerting national authorities to the disaster, the ship's initial signals were addressed to naval headquarters, and the news was quickly passed on to other authorities. The first 'all stations' call was broadcast by MV *Northumberland* at 11.20 a.m.[29] The *Taranaki* sent a message three minutes later, which was received in Wellington. By this time, independently, Gisborne 'ham' operator C.T.C. Hands had told an Australian contact there had been a major earthquake in Poverty Bay, though he was unaware of the scope of the disaster further south.

Just after noon, Captain Upton of the *Taranaki* suggested 'getting decent water then asking [the] *Veronica* if we can assist'. Fortunately the bow was pointing out to sea. There was so little water beneath the *Taranaki*'s keel that the propellers churned through mud, and she bumped against the bottom several times until, about three miles (5 km) offshore, she finally floated clear. Captain Wood of the *Northumberland* advised his employers at 12.29 p.m.:

> GNRF [*Northumberland*] to ZLW SAVILLINE, WELLINGTON — After earthquake water at anchorage fell 10 feet [3 m] proceeded to sea and waiting in vicinity. Napier badly damaged severe fires broken out. Harbour apparently blocked. TARANAKI suffered severe shocks trembling and bucking violently — Wood.

By this time the newly arrived merchant *Waipiata* was also calling for help on shortwave — interfering with the *Veronica*'s attempts to do the same. At 12.50 p.m. — by the *Taranaki*'s radio log — the *Veronica* sent a formal general distress signal by morse, broadcast to all possible listeners, each word repeated twice:

> GLUD to CQ: Serious earthquake at NAPIER. All communications destroyed. Medical assistance urgently required.[30]

A few Napier ham operators had their stations running by early afternoon. G.E. Tyler, operating 2GE by battery from his home in Napier South's Vigor Brown Street, contacted others in the South Island. Around 5.00 p.m. he got in touch with Wellington ham stations 2GK and 2BI. One of these stations, run by S. Perkin, was subsequently requisitioned by the Post Office for official purposes.

Indirect telephone contact was established within a few minutes of the quake through the Public Works telephone system, a network laid parallel to the main power transmission lines and quite separate from the Post Office system. Linesmen working eleven miles (18 km) south of Taradale were able to call Mangahao power station from a lineside hut. They were unable to call Redclyffe, however, and drove back to the substation to see what was going on. By mid-afternoon they had returned to the hut to report the damage to Mangahao.[31]

Cabinet met early in the afternoon to discuss the emergency. Early news did not seem good, and Minister of Lands R.E. Ransom and Minister of Health A.J. Stalworthy left by train at 2.00 p.m. for Palmerston North, intending to drive to Hawke's Bay that night to coordinate relief efforts. However, news from the disaster zone remained scanty, and shortly after 2.00 p.m., Prime Minister George Forbes signalled Commander Morgan directly, anxious for 'any information as to extent medical assistance required'. Morgan's terse response seemed alarming: 'As much medical assistance as possible. Whole town wrecked and fires raging.'[32] This dramatic picture was further detailed in staccato tones by the *Northumberland*, direct to Forbes, a few minutes later:

> Heavy tremors still continuing, fire raging Beach Road [sic] Hastings Street district looks as if all houses lost fire also Harbour reserve raging furiously Bluff Hill fallen up to flagstaff burying road several houses demolished by shock. Doctors Nurses urgently required ships *Taranaki* and self launching relief parties Doctor and hospital trained attendants. Freshening NE wind driving flames over town.[33]

Just after 3.00 p.m., Commander Morgan advised that it was 'impossssible' to estimate the extent of damage but 'it is very severe. Have taken charge and am endeavouring to organise situation ashore. Have assistance of SS *Taranaki* and *Northumberland* every available man landed and refugees are coming on board *Veronica*. Shall remain in inner harbour and in touch with situation ashore — commander, *Veronica*.'[34] A further signal at 4.30 p.m. dissipated any thought that the disaster might have been restricted to a single town:

> Am informed that Hastings and Waipawa and Waipukurau have suffered equally with Napier (stop) Medical assistance is urgently required there and organisation for food etc in all towns (stop) Am endeavouring to do this at Napier, but assistance is urgently required elsewhere — Commander HMS *Veronica*.[35]

For Cabinet this was the worst possible news; the quake was no local shake but a major regional catastrophe. It had nationwide implications. Minister of Defence J.G. Cobbe and the Honourable R. Masters left Wellington at 6.00 p.m. by car. Before leaving, Cobbe directed the Army to organise a relief convoy from Trentham. Chief Inspector of Explosives R. Girling-Butcher was intercepted on the train for New Plymouth and asked to drive to Napier. Forbes, meanwhile, appealed for doctors and nurses to hasten into the disaster zone. Thirty-eight doctors and 75 nurses had gone by the end of the day.[36]

THE DESTRUCTION OF NAPIER

Fire continued to rage in central Napier through the afternoon. In the face of flames driven by a brisk breeze, and frustrated by lack of water, the firemen fell back time and again — but they never gave up. Loss of the water supply was a major problem. It was not wholly due to cracked pipes: the real difficulty was that power failure prevented the borough pumps from refilling the Cameron Road reservoir. Once the water in the reservoir and pipes had gone there was no more. Emergency power was unavailable — the earthquake had done serious mischief to the borough plant, which had been kept as a standby system after

Napier fire damage, February 1931.
Source: 'Council of Fire and Accident Associations of New Zealand. Official records of Napier Earthquake, February 3rd–10th 1931'. Also sources cited in text.

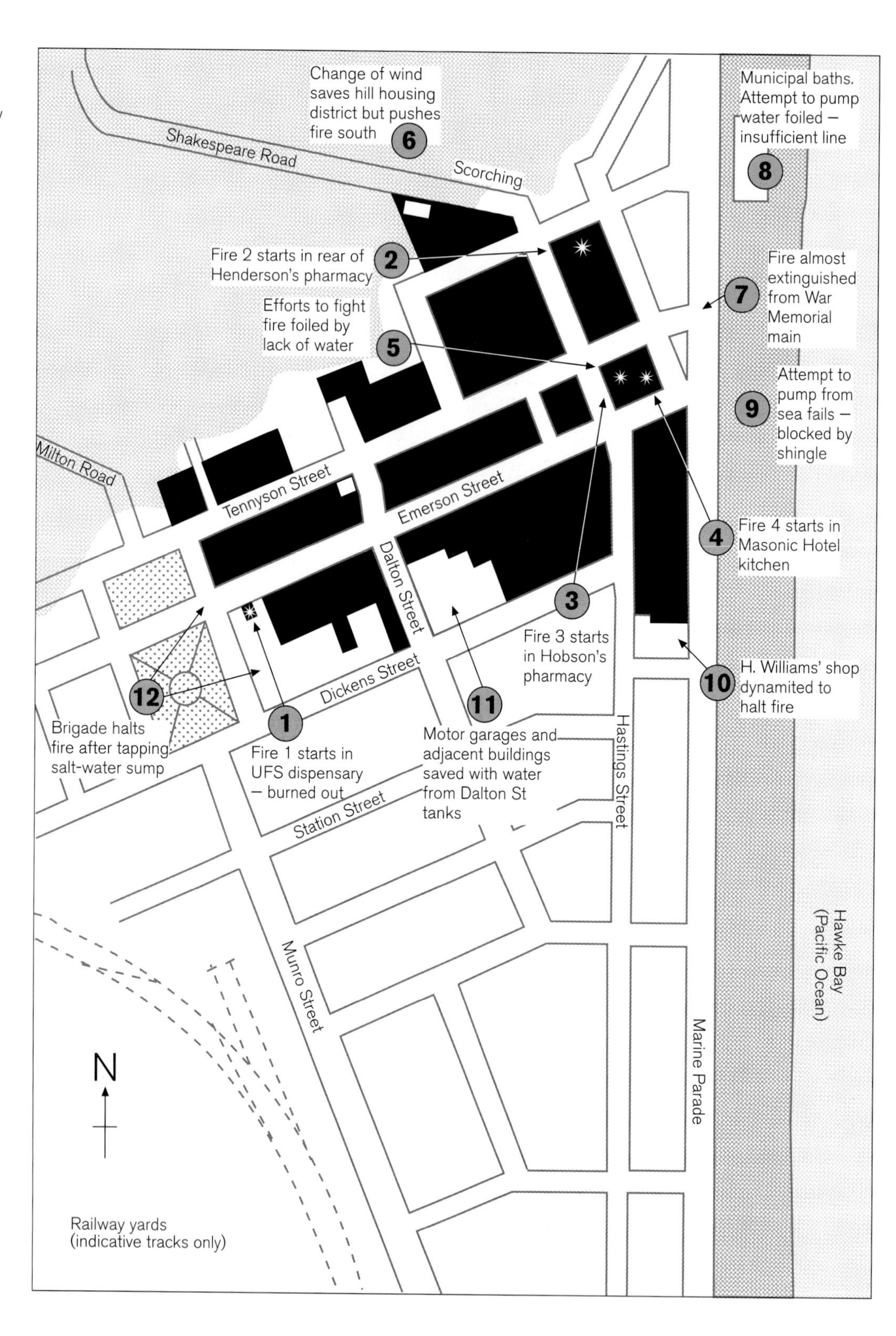

national grid power had arrived in the late 1920s. The three producer-gas engines were wrecked and all the lead-acid batteries had toppled from their stands. The two diesels were intact, but circuits could not be made because the transformers had been thrown from their tracks and lost their oil. There was no chance of restoring power quickly. Napier seemed doomed to burn.

The brigade had been unable to halt the fire as it advanced up Hastings Street towards the vulnerable housing district on the hill, though they worked on until the last moment and were lucky to save the engines as fire rolled over their position. Flames licked into Shakespeare Road, singeing the front of the Clarendon Hotel and adjacent structures such as the Humphries and Humphries building. Some houses on the lower slopes were gutted, though curiously other downhill structures were only damaged, while others further up the street were burned out.[37] Flying cinders ignited tinder-dry gardens further up the hill. Even with the help of sailors from the *Veronica* there seemed little chance of halting the fire until a change of wind around noon blew the flames the other way, back down Hastings Street towards the southern residential district.[38]

Still searching for water, Gilberd took one of the engines to the fire main next to the Soldiers Memorial on Marine Parade. The brigade managed to get enough pressure here to temporarily stop the fire spreading down Hastings Street, but the supply ran out before the blaze could be completely extinguished. Gilberd then had a hose run from the municipal baths on the Marine Parade beach, in an effort to stop the fire reaching Henry Williams' premises, but this was 1500 feet away (460 metres) and he had only enough gear for a single lead. This couldn't deliver enough water, so Gilberd decided to draw from the sea closer to the fire, and had an engine backed down into the surf over planks hastily laid by willing volunteers.[39]

By this time the fires were spreading in two directions — south towards Dickens Street, and southwest through Tennyson and Emerson Streets towards Clive Square. Around 3.30 p.m., when Bill Ashcroft arrived to look for his father, the blaze had reached the Murray Roberts building on the corner of Emerson Street and the Marine Parade. The situation was critical. Hampered by shingle blocking the pump intakes, Gilberd soon abandoned efforts to use sea water and sent the No. 2 Dennis to Clive Square, where firemen dropped lines down a half-forgotten salt-water sump dug decades earlier and began battling the fires ripping down Emerson Street.

Meanwhile Gilberd took the No. 1 Dennis to Dalton Street, where he drew on storage tanks in the main pumping station and began fighting flames in the Caledonian Hotel, on the corner of Hastings and Dickens Streets. On the other side of Hastings Street, the fire had reached the Napier Dairy Company's No. 1 Store and looked set to roar past Albion Street into the residential district. The

Dalton Street water — and a timely demolition charge that flattened Henry Williams' shop — was just enough to halt the fire before it could leap into the wooden housing district south of the town centre. Humphries Cash Groceries, near the Albion Street corner, was one of the last Hastings Street shops to be destroyed, evidently around 5.00 p.m.[40]

Gilberd was also able to stop the blaze advancing southwest through the Dickens Street motor garages. Anderson and Hansen's garage was gutted, but the brigade was able to save the adjacent Stewart Nash Motors, the Hawke's Bay Rubber Tyre Fusing Company, and W.D. Dalley's garage with his vulnerable petrol bowsers.

While Gilberd led the fire-fighting effort in Dickens Street, firemen working with the No. 2 Dennis at the Clive Square salt-water sump were able to halt the fire at Kelly's Cash Store, in Lower Emerson Street on the seaward side of the previously incinerated Friendly Society pharmacy. They also saved a few structures on the border of the square and on the corner of Dickens Street, including Whitfield's Motor Service on the corner of Emerson Street and Clive Square. Even so, the struggle was touch-and-go at times. 'When the water supply ran out,' a reporter later wrote, 'the fire brigade devoted itself to pulling down and dynamiting, so as to keep the fires within bounds.'[41] The power of the fire was enormous. Smoke gushed thousands of feet into the sky. Refugees gathering in Nelson Park, half a mile away, were battered by 'deafening explosions' and bombarded by 'bits of iron landing at our feet'.[42]

The firemen and the many public volunteers who joined the fire-fighting effort could not relax even when a southerly change of wind at dusk seemed to reduce the risk of the blaze spreading. Napier's business district from Tennyson to Dickens Streets had to be abandoned to the flames, which continued to dance and roar for another eighteen hours. 'There are constant detonations,' a Wellington reporter who arrived that night wrote, 'which some people think are earthquake sounds, and others think are falling walls, but which may be new dynamitings by the firefighters.'[43] Nor was the danger over. The wind swung to a northerly during the night and began pushing the fires due south over the Dickens Street block. The brigade fought on, some of the men collapsing from exhaustion,[44] without being able to extinguish the blazes. The Power Board store caught alight as late as 6.00 a.m., and there were real fears of a further crisis. This was averted, but virtually the whole of central Napier was annihilated — more than eleven blocks were either totally or partially burned out.

Fire also roared through parts of Ahuriri on the other side of the hill. Water was again a problem, compounded by the fact that the main brigade was busy in town. Geography helped. The fire started in the upper floor of the Robjohns Hindmarsh building on the eastern spit, which was bounded on two sides

In Napier a change of wind around noon blew the fire away from the hill, saving the vulnerable residential district but pushing the fires back through the town centre. Here, smoke blots out the sun as fire roars along Emerson Street, just two hours or so earlier a bustling main thoroughfare. The blaze completed the destruction left undone by the quake, and the hard-pressed brigade was able to stop it only at Clive Square.
(Storkey Collection)

The fire advancing down Hastings Street looked almost like the burst of a demolition charge, rolling away from the town centre and engulfing everything in its path. The new post office is visible on the left; it was only slightly damaged by the quake, and despite being gutted by the fire was later refurbished. The tram tracks are noteworthy: the service opened with great fanfare in 1913 but was never reinstated after the quake, and the trams ended their life ignominiously converted to huts in the Kennedy Park motor camp.
(Storkey Collection)

Moments later the fire has jumped to the next buildings. Wooden Napier, the remnant of old settler days, largely survived the shock but was then destroyed by flame. The fire has almost reached its furthest advance. It was stopped a short distance to the right of the photographer by a timely demolition charge and with the help of water from the Dalton Street storage tanks.
(Storkey Collection)

Havelock North apiarist Bill Ashcroft drove into Napier to look for his father W.H. Ashcroft. He arrived about 3.15–3.30 p.m. just as the fire reached its height.

'Napier was burning and looked worse as I approached. From miles away you could see the smoke and flames streaming inland on a strong sea breeze. As I went up the Parade I passed people carting their furniture from their houses to the beach which was crowded with people surrounded by their possessions. Everyone seemed quite cheerful and curiously indifferent. All the way along brick fronts had fallen out exposing the interiors of rooms. I stopped the car about a block from the Masonic Hotel and walked on.

'I simply couldn't believe my eyes. The fire was just starting on Murray Roberts at the corner of Emerson Street and the Parade. The Masonic Hotel was a blazing ruin. Looking down Emerson Street towards R & G's all you could see through the smoke was piles of bricks with a few buildings standing here and there. It was no good looking for Dad THERE. For all I knew the entire business population of Napier was under the bricks, so I carried on along the Parade looking for someone I knew. The fire brigade were pumping water from the sea and one fireman said the only casualties were at the Tobacco Company at Port where 6 girls were killed. I knew that couldn't be true.

'I took the car back to Georges Drive to try and find Morice in his home. Here the ground was covered with huge cracks up to a foot [30 cm] wide and three or four feet [a metre] deep. Houses were leaning at all angles where the foundations had given and blocks of houses had been shifted back a few feet. I met Benny West. He told me that the High School was badly damaged

by the sea and on a third by the Iron Pot dockway. The North Pond beyond — a remnant of the old swamp from colonial days — formed a wider barrier between Ahuriri and the hill. As a result the blazes did not wreak such widespread destruction as in town, but they were intense enough, partially fuelled by 400 casks of tallow stored in the Vestys warehouse.[46] Flame engulfed the post office, wool warehouses and stores depots, and, ominously, approached the oil tanks on Hardinge Road. The port brigade pumped water from the Iron Pot, stopping the fire from reaching the tanks, but again could do no more than fight a containing action and the fires were still roaring away the following morning. Fortunately the residential district along Hardinge Road and Waghorne Street was saved.[47]

Weakened walls collapsed during the afternoon of 3 February as aftershocks pounded the town. Gerhard Husheer's vigil in Dr Moore's precariously standing hospital ended when Mervyn Barggren, Ernest Barr and Alfred Hopewell arrived. Barggren had been in jail awaiting trial for burglary, and the other two had been undergoing questioning by police over alleged vandalism when the earthquake hit. After the trembling subsided all three were released to assist. Barggren heard that a patient had been trapped inside Moore's hospital and climbed the front of the building to reach him. With block and tackle the three lowered Husheer to the ground, and he was taken to the beach

but no-one hurt.... He knew nothing of the offices in town. The thing had happened so suddenly that no-one knew anything except their own particular building. I managed to get along Georges Drive by going on the grass and the footpath sometimes and went up to the hospital. A crushed car on the way up prepared me a bit, but I never thought to see such a scene as met my eyes at the top. The Nurses' Home forming the entrance block and gateway was absolutely flat, what was originally a two storey brick building being a heap of bricks four feet high. A tractor was at work pulling lumps of masonry away, and from the casual way they were working I didn't realise that nurses were under that pile, but about six night nurses were killed there. The hospital buildings were still up but with gaps all over where walls and roofs had fallen in. Looking down on [the] Port the buildings there could be seen in ruins while a fierce fire was raging. Tied up at the wharf was the sloop *Veronica* and men from her were working at the hospital.

'I worked for a time up there putting salvage gear on a lorry and then met a chap from Sargoods who said the whole of R & G's staff had got out safely. So I went on to the Parade again in the hope of seeing Dad. By this time the fire had spread from the corner and Henry Williams' iron-mongers store and the Majestic Theatre were burning fiercely. There seemed every prospect of the fire sweeping the whole area as the water failed with the quake. Not seeing Dad I went home to find him there having arrived just after I left.'

W.J.C. (Bill) Ashcroft[45]

The change of wind about noon blew the Ahuriri fires towards the Hardinge Road residential district and sent smoke billowing over the hill.
(Storkey Collection)

with the other patients from the hospital. Barggren went on to rescue a number of others and volunteered his services as a cook in the Nelson Park refugee camp — acts that earned him the respect of the courts when he was sentenced a month later.

POWER AND WATER — HASTINGS

Hastings also faced destruction. Fire broke out near Roach's devastated department store soon after the quake. The brigade was hampered by the collapse of the fire station, on the corner of Market Street and Lyndon Road. It took about 30 minutes to move the debris trapping the Dennis fire appliance, with the help of a truck from the neighbouring Williams and Creigh yard, and 'by skilful driving the engines climbed over the wreckage without capsizing'. As in Napier water was a problem, this time because the road bridge over the old Ngaruroro riverbed had collapsed, taking the main with it and cutting the link to the reservoir in the Havelock hills. The four wells supplying the system from the aquifer under Hastings were still available, but — again as in Napier — power failure rendered the pumps inoperative. A fire in the Union Bank building was put out with all available water from Tong's artesian well, but firemen could not 'impose a check on the continuing onward swirl of fire'.[48] Rescuers rushed to save trapped victims. Hugh File was pulled clear of the rubble that had been Roach's department store just in the nick of time.[49]

Hastings was apparently doomed, but the town — nicknamed the 'city of blazes' during its boisterous youth — was saved by the municipal electrical engineers. Power had been supplied for many years from diesel-driven generators operating in the purpose-built Power House on Eastbourne Street. The system was mothballed in the late 1920s when the town was connected to the national grid. The Power House — which remained the hub of the distribution network — was badly damaged in the quake, but the plant remained intact and the engineers had the big General Electric diesels running within an hour and a half. They sounded 'a little noisy', suggesting that 'the heavy foundations ... had been loosened',[50] though that was a minor concern in the circumstances. Linesmen began restoring broken connections, prioritising power for the hospital and water pumps. Cables were even run directly over rubble to a nearby B-class radio transmitter. It proved inoperable at first, but by this time the first calls for help had been made from Ahuriri.[51] Later the station was used to contact Gisborne.

Hospital services were quickly overwhelmed. Hastings relied mainly on the Napier hospital and had only two small establishments of its own: the public

War Memorial Hospital, which had opened — after much lobbying — in the late 1920s, and Royston private hospital. Both were damaged, and in any case neither could have handled the flood of injured. Hospital doctors quickly resurrected plans first implemented during the 1918 influenza epidemic, and established an emergency centre in the racecourse tea kiosk, supervised by Sister Edith Williams. Among those on site were Dr A.D.S. Whyte, who 'worked in the operating room for hours after he knew his daughter had been killed'. She had been buried under tons of rubble while having her hair done.[52] Many people volunteered to help, Reginald Gardiner and Bill Whitlock of the *Herald Tribune* among them — Gardiner did not return home to Havelock North until late afternoon, his appearance bringing 'great joy and relief' to his family.[53]

Much of this effort was improvised, but the organisation later drew the admiration of Christchurch engineer E.F. Scott. Casualties were examined by a doctor on arrival and ticketed 'theatre', 'ward' or 'outpatient'. Their details were taken by a minister for relaying to the YMCA information centre. Some were given morphine —

ABOVE: Grand Hotel proprietor J.A. Ross was trapped in the cellar by the quake when his building was 'rent in twain'. He was thought to be safe until fire ripped through the fallen building during the night and 'consumed what remained of the building'. Also visible here are the ruined premises of H.O. Shattky and Henry K. Thomson.
(S. BARKER LOAN, ALEXANDER TURNBULL LIBRARY F-139759-1/2)

LEFT: Hastings suffered scarcely less than Napier during the quake. Racked and twisted roofs couple with the remains of the Bank of New South Wales in this depressing scene.
(S.C. SMITH COLLECTION, PACOLL-3082, ALEXANDER TURNBULL LIBRARY 47690-1/2)

The 'gully of destruction' – Heretaunga Street from the corner of Nelson Street, looking towards the Havelock hills. This appears to have been taken some time after the quake, by which time much of the rubble had been swept aside. The amount of debris is still staggering.
(S.C. SMITH COLLECTION, PACOLL-3082, ALEXANDER TURNBULL LIBRARY G-47693-1/2)

indicated by a piece of sticking plaster on their forehead. Equipment came from diverse sources, mostly dentists — some brought gas cylinders to the raceway and two arrived with portable X-ray systems. Chemists hurried to the ruins of their shops in the hope of finding useful drugs. However, there were serious shortages of other medical stores, particularly gauze pads. Volunteers sewed makeshift pads.[54] Some 2500 patients were treated on that first day, and 66 operations were performed with sterilisation and anaesthetics.[55] Torches and lamps had to be used after 9.00 p.m. when a particularly large aftershock damaged the Power House, forcing the engineers to close down the diesels until shoring could be put in place.

The hospital retained a makeshift look despite the efficiency of its staff. Bill Ashcroft took his sister Dorothy — a registered nurse — into Hastings on 4 February to help, after she had worked overnight in the Greenmeadows facility.

> The place looked like old paintings of a sacked city with no lights and people gathered around open fires. While I was there there was a shock and the building swayed and creaked and groaned but nothing happened. The next week the patients refused to stay in the building as they considered it unsafe, so the hospital closed down and the patients were sent to Palmerston North.[56]

The grisly task of dealing with the dead began an hour after the quake. Dr Cashmore worked with police to set up a mortuary and receiving depot in the Hastings YMCA buildings under Sister H. Dillon. By the end of the day the centre had handled 50 bodies. Every effort was made to find the dead — E.F.

Auckland pipe manufacturers Stewarts and Lloyds wrote to the Havelock North Town Board after the quake, wondering how the pipes they supplied in 1914 had generally stood up to the shock. A photograph in the *Auckland Weekly News* of the fallen bridge over the Karamu Stream, with its single surviving pipe, particularly drew their interest. 'We should be glad to know whether this is our 9" steel pipe,' they wrote.
(HNL 0161, HAVELOCK NORTH PUBLIC LIBRARY)

Scott remarked that even the 'bones of carcases recovered from the ruins of butchers' shops, and dressed wax models from drapery establishments' were brought in. Some bodies were 'beyond identification', and a few, 'through certain injuries', could not be placed inside. Everything was recorded, 'including measurements down to size of foot, and samples of clothing and all trinkets etc.'[57] The *Herald Tribune* later expressed public thanks for the 'magnificent work' of the undertakers. 'Words fail to express the fine depth of courtesy, gentleness and reverence shown by the police in the handling of the dead.'[58]

The Borough Council quickly rallied. Although Mayor G.F. Roach was away at Waikaremoana, Deputy Mayor Robert Henderson, who was also the Fire Brigade Superintendent, instructed Town Clerk Percy Purser to convene a public meeting around 2.00 p.m. outside the Bank of New South Wales. Quick decisions followed.

> First and foremost, all military men were asked to meet in half an hour at a certain spot, elect a head of committee, and to promptly clear, picket and then police the damaged area ... All carpenters and builders were asked to meet at another spot, elect a committee, and organise the systematic recovery of the injured and dead, and under direction of the Borough Engineer, pull over all dangerous walls ... Lawyers, bankers, businessmen and other leading citizens were then asked to elect a Hastings Borough Earthquake Executive Committee.[59]

Local solicitor and former officer Colonel H.H. Holderness was appointed chair of this 'citizens' committee', which Scott praised as 'the key of the whole situation … responsible for the splendid results obtained'. Holderness had led a local battalion during the war and was able to collect many of his former soldiers. He appointed high school principal Major W.A.G. Penlington to run a 'force of 150 special police' in six-hour shifts, picketing the town and preventing the 'valuable contents of shops, warehouses and businesses' from falling victim to 'the light-fingered and the pilferer'.[60] They divided themselves into groups of three, armed with pick and axe handles and wearing armlets and belts for identification. By about 4.00 p.m. they had cleared all but authorised personnel from the centre of Hastings.

Henderson was finally able to convene a Hastings Borough Council meeting in the early evening outside the municipal buildings. The electrical engineer was 'authorised and instructed' to make essential repairs and get the street lighting working. The borough engineer received authority to hire 40 men to dig graves, and to hire additional labour to restore the water services. Other priorities included street clearing and getting water to the emergency hospital.[61]

The New Zealand Permanent Air Force (NZPAF) — forerunner of the RNZAF — flew into Bridge Pa aerodrome near Hastings mid-afternoon and began ferrying doctors and emergency equipment, an effort coordinated by the *Veronica*.[62] The only other airfield in the district, the Beacons aerodrome near Napier, had been damaged by the quake, though aircraft later put down there with care. The biggest problem for the NZPAF was resources: they had only a few aircraft, which seriously restricted the scale of assistance they were able to give. The Auckland Aero Club quickly put its own machines at the disposal of the military. Five aircraft left Hobsonville during the afternoon of 4 February to bring a donated chlorinating plant to Napier.[63] Later the NZPAF led the aerial reconnaissance of the district.

AROUND HAWKE'S BAY

In Havelock North there was less damage than elsewhere — partly because there was less to smash — and no deaths, though some Havelock North people died in Napier and Hastings. Soon after the shake Bill Ashcroft left for Napier to find his father. As he drove through Havelock North township he found:

> … groups of people … standing about in awed silence while over the brow of the hill the school kiddies were coming home scared, some crying. Minor

> shocks were continuing ... In the road motors and lorries were coming hell-for-leather with husbands coming home or parents coming for their children. To add to the confusion a mob of sheep panic-stricken were milling around in the road dashing from side to side as the cars tore through them.[64]

He was unable to find his father, but the elder Ashcroft hitched back to Havelock North and arrived home later in the day.

At Waimarama, the sight of a new reef just offshore rang alarm bells in the minds of Dorothy Campbell and her family.

> If the sea had receded the chances were that a tidal wave might be expected ... Bro. Frank who had managed to join us by then favoured the latter [land had come up], so did I but as I certainly was not sure I ran the car out & heated it up ready to start at a moment's notice. All this time as you can imagine I was frantic about Neil, in fact about all the family as I knew that earthquakes were usually never as bad on the coast as they were inland & it was bad at Waimarama ... Neil arrived in 1½ hours with two men to clear the road to say all were safe but Horonui was terribly badly damaged & his brother's place had collapsed & he could see that Napier and Hastings were on fire. By that time we decided we were safe from tidal waves & Neil told us to stay there as he thought his family would all come out.[65]

For Ernest St Clair Haydon the first priority was taking 'poor old Mrs Daly' to a doctor — her arm had been crushed by a falling chimney. Haydon set up a tent in his orchard and hastened into Napier to see what he could do, helping an insurance company manager retrieve documents from wreckage before going to the nurses' home to see who he could pull out of the debris there.[66]

At Mohaka, J.R. Murphy and his assistant found that their horses had bolted, so they walked back to the homestead. It had been twisted and tipped from its piles, and the 'whole household' were outside. They had already taken bedding, food and equipment out of the house and prepared to camp in the home paddock. Later they were joined by other refugees from the district.[67]

Herbert Guthrie-Smith returned to his Tutira home to find a puzzle. There was not a window broken, yet chimneys had:

> ... snapped off at roof level, leaving huge gaps in the iron. Tanks crumpled and torn lay beside their stands. Into the air water pipes vaguely protruded. Within the house itself what the heavier furniture may have been about could only be inferred from indentations on the wall surfaces. Bricks, brick dust and mortar littered the floors and passages. Mantlepieces were swinging loose. Doors

> were tight jammed or would not close. Pictures, though with here and there an exception, had been clean flung from their fastenings, or on their twisted strings hung awry and face to wall. Books, splintered glass, and broken china strewed the carpets. Except in the case of inkpots, it was impossible to say what may have been the movements of any particular article.[68]

To Guthrie-Smith's incisive scientific mind the wreckage posed mystery as much as a cleaning job. A piece of paper had jammed under the putty of one window, 'demonstrating what weird, unwonted air-pressures must have been circulating in that demented room'. Amid the jumble of his sitting-room a cigar box sat where he had left it on a bookshelf, yet one cigar inside had been split in two. A section of the central chimney had smashed through the floor of the verandah — but not the roof, though it should have hit that first. Guthrie-Smith found the chunk on the ground beyond, where it had apparently bounced over the intact verandah rail. 'Dare we suppose the house, first flat on its face and then flat on its back?' he conjectured.[69]

THE LONG NIGHT

People from Wairoa to Waipukurau settled down to spend the night outside. Some had tents, others made do with blankets and rugs. Napier jeweller L.S. McClurg arrived with his family and neighbours at his friend Ernest St Clair Haydon's Puketapu farm, where they slept 'in motor cars or under the trees'.[70] Fortunately it was fine and warm — 'Never did a softer night smile upon human catastrophe,' a reporter later remarked — but few people actually slept. Many camped on the Marine Parade, where journalists rushing up from Wellington found them 'sleeping out on the seashore amid stacks of hastily salvaged goods ... These refugees knew, or had been told of, the lifting of the coastline. But here they were, seeking slumber on the sands [sic] between a line of burning buildings and the line of surf.'[71]

Some were able to improvise a meal. Mary Hunter and her husband gathered with friends at a house on top of the Napier hill, where they had a 'lovely hot boiled potato' around 7.00 p.m. Their two hosts were waiting on news of their daughters. 'They were so brave and unselfish in their anxiety,' Hunter later wrote.[72] Both daughters had been killed.

K.C. Sinclair, on Napier Terrace near the ruins of the hospital, was worried by aftershocks:

> Most of the neighbours had gathered on our lawn and had brought bedding with them, and we were busy making beds ... fortunately it was a beautiful moonlit night ... It was, in spite of our terrible plight, really amazing to see the way people behaved. One man had a cap on, another wore a hat, as did several women, to sleep in. I myself had in the morning just put on a dress to attend a wedding and have worn it since, as I have lost everything else. No-one, of course, undressed ... In the middle of the night we heard an exclamation from one woman. She had trod on something prickly — it was a hedgehog. Everybody laughed — and then came another terrific shock, which set our nerves all on edge again.[73]

This aftershock was a major earthquake in its own right, damaging buildings already weakened by the primary shock, including the Hastings Power House, where the engineers had to shut down the generators and find more timber to shore up the roof. Loss of power cut off the water, and fire raged through the ruins of the Grand Hotel, killing proprietor J.A. Ross who was still trapped in the basement. The same shock was felt in Wairoa, where ham wireless operator R. White was on the air at that moment. 'I am hanging on,' he signalled dramatically. 'Houses are falling all round.'[74]

Other people worked through the night to restore communications; a 'shaky circuit' brought Hastings back into telegraph contact with Wellington around 1.30 a.m. on 4 February.[75]

Fires in both Napier and Hastings lit the sky with an eerie red glow. From Te Mata, Bernard Chambers recorded that 'the glare of both Hstgs & Napier on fire lit up the sky'.[76] Bill Ashcroft, in Havelock North, saw flames rising 'hundreds of feet' through the night.[77] Rescuers driving up from the south saw the loom of the fires from even further away, while from the merchantmen out at sea Napier 'looked like a vast furnace, flames and smoke sweeping from the promenade into the centre of town'.[78] In Napier itself, Mary Hunter found that the 'night was windy and the dust and smoke from the fires in the town was very disagreeable and made us all filthy'.[79]

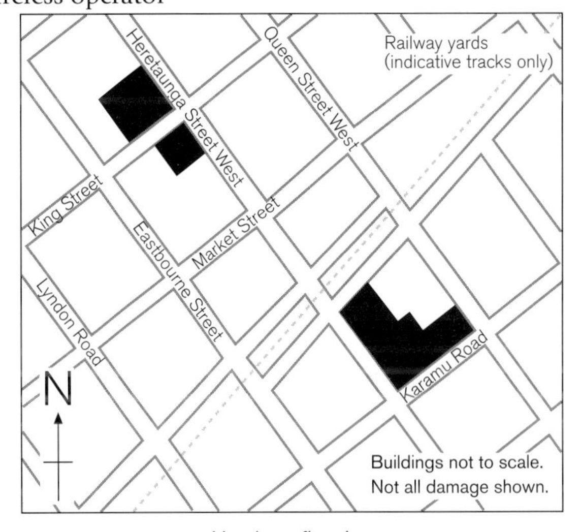

Hastings fire damage.
SOURCE: 'COUNCIL OF FIRE AND ACCIDENT ASSOCIATIONS OF NEW ZEALAND. OFFICIAL RECORDS OF NAPIER EARTHQUAKE, FEBRUARY 3RD–10TH 1931'. ALSO SOURCES CITED IN TEXT.

CHAPTER 4

'Thank God for the Navy!'[1]

At dawn on 4 February the cruisers of the New Zealand Naval Division, HMS *Dunedin* and HMS *Diomede*, arrived off the Napier roadstead. Wary after reports of shoaling offshore, they slowed well out to sea and cautiously worked their way in. By 8.30 a.m. they were at anchor about two miles from the harbour entrance.

Their arrival in such short order was the result of a turn of fortune. On the morning of 3 February they had been about to leave Devonport for manoeuvres with the Royal Australian Navy. They were fully manned, supplied and fuelled, and had steam up when Morgan's SOS came in. After a second signal at 11.27 a.m. revealed the 'precariousness of the situation' in Napier,[2] Commodore Geoffrey Blake cancelled the manoeuvres and contacted Dr C.E. Maguire, Superintendent of Auckland Hospital, asking for doctors, nurses and medical supplies. The response was a 'genuine triumph of organisation'.[3] Eleven doctors — including Dr H.L. Gould, the Assistant Medical Superintendent — and seventeen nurses arrived within 90 minutes. They were joined by the Reverend G.T. Robson, MC, chaplain of the shore base HMS *Philomel*. Surgical and medical stores, an X-ray plant provided by Philips Lamps, 54 stretchers, 5 marquees, 34 tents, 400 blankets, 125 beds, 200 ground sheets, 80 shovels and 21 picks were collected on the dockside by 2.00 p.m. and loaded within half an hour. The two cruisers immediately put to sea, working up to 24 knots and rounding North Head by 3.00 p.m.[4]

Both cruisers were able to maintain their speed through the night, which brought them to Napier the following morning. Few on board slept. Crews were divided into rescue parties during the voyage and given training in stretcher-bearing and first aid.[5] Galley hands worked to bake bread for distribution to the earthquake victims, while others set up the portable X-ray machinery in the *Dunedin*'s torpedo flat.[6] Each cruiser brought 450 officers and men to the disaster zone — not all could leave the ships, but with the *Veronica*'s crew they made all the difference. Nor were the cruisers the only aid the Navy could bring to bear if necessary. The *Veronica*'s sister ship HMS *Laburnum* was at Whangarei. Blake ordered her to Auckland, where she refuelled and loaded medical supplies in case they were called on.[7] Australian Prime Minister J.H. Scullin also offered the services of a Royal Australian Navy cruiser then in Tasmanian waters.[8]

The Napier fires were still spreading when the Navy arrived — Ahuriri continued to blaze, and in town the Power Board store off Dickens Street caught light at 6.00 a.m.[9] The situation was so alarming that serious consideration was given to sending in naval demolition parties to create fire breaks. Blake, who had gone ashore, signalled HMS *Dunedin* to 'send the lieutenant in charge of the torpedo stores, with all available demolition stores, to report to the *Veronica*'.[10] Fortunately the brigade, naval 'bluejackets' and others were able to keep the fire from spreading into the residential districts; the Power Board store was the last major structure to catch alight, and as the day progressed the firefighters were able at last to get on top of the situation.

The Navy made an enormous contribution over the next few days. One observer suggested that there was 'without doubt, an entire absence of organised effort until the Navy arrived'.[11] This was not actually true. Police, fire brigade and medical services played an important part, as did many individuals who survived the quake. However, the influx of hundreds of fresh, disciplined men who were not facing the loss of all they knew made a real difference. The marines set up headquarters at the Byron Street Police Station under Royal Marine Captain Hardy Spicer of HMS *Dunedin*. 'The first two days were the worst,' Spicer later recalled, 'for the men had no sleep whatever.' Royal Marine

Commodore Geoffrey Blake (1882–1968), Commodore in Command of New Zealand Station 1929–31, coordinated the naval relief effort in the wake of the earthquake.
(Alexander Turnbull Library F-87921-1/2)

HMS *Dunedin*. Her raised 'trawler' bows are particularly evident in this picture.
(S.C. Smith Collection, PAColl-3082, Alexander Turnbull Library G-47352-1/2)

Captain J.C. Westall of the *Diomede* later explained to reporters that they 'acted as firemen, took part in demolition work, cooked meals for refugees, were food distributors, water carriers, tractor drivers, policemen, searched for the dead and carried bodies to the morgues, carried out sanitation work, and even acted as nurses in the hospitals'.[12]

Both cruisers moved closer to Napier on 5 February. Their navigating officers, Lieutenant Commanders Harper and Clarke, spent 4 February carrying out a thorough survey of the harbour approaches. They found that the bottom had been lifted generally by five feet (1.5 metres), and with this information were able to bring the cruisers a mile closer inshore.[13]

The Army also arrived on the morning of 4 February, reaching Napier slightly before the Navy. The convoy requested by the Minister of Defence had left Trentham under Major Nicholls late on 3 February, comprising four officers and 27 men with 500 bell tents, 12,680 blankets, and equipment to set up a field kitchen able to cater for at least 1000 people. They left Thorndon railway station on a fifteen-car relief train, expecting to be dropped off at Ormondville. En route they were advised that they could get to Waipukurau. Nicholls asked for trucks to meet them when the train arrived at 1.30 a.m. on 4 February. The equipment was trans-shipped in two hours, and the convoy reached Napier railway station at 6.15 a.m. Here they awaited instruction from the emergency committee meeting due to be held at 7.30 a.m. Around 8.00 a.m., after some confusion, the Army was instructed by Hawke's Bay Hospital Board Chairman C.O. Morse to set up the equipment in Nelson Park.[14]

Ruined buildings in Napier, a day or two after the quake.
(Johnson Collection)

THE MORNING AFTER — WEDNESDAY, 4 FEBRUARY

Napier presented a dismal scene on the morning after the quake. 'At Clive Square there were ruins everywhere,' one witness commented. 'Looking up Emerson Street all one saw was a vast ruin, with people everywhere looking at the destruction; it was an amazing sight. Fires were everywhere.'[15] The town centre continued to smoulder, while naval ratings and locals got to work demolishing unsafe buildings and searching the ruins. They were joined by others from the surrounding district. 'Of course, at peep of day all the men available were back in Napier to try and help,' Ernest St Clair Haydon later wrote. 'Some of the sights were too appalling for words. We were still getting some very bad shocks and practically speaking a continual tremor.'[16]

Police working through the night recovered 40 bodies by 6.00 a.m. on the 4th, and were later joined by the bluejackets.[17] A visiting doctor saw the naval parties 'clearing a lane through Hastings Street' and later found a group:

> ... trying to pry an enormous mass of masonry from something. It is smouldering all round. The bricks are too hot to touch with the naked hand. The sweet, nauseating smell of burning flesh ... reaches the nostrils. Some unfortunate has been caught there. They are trying with pick, crowbar and lever to get out the poor human remains. It seems as hopeless as lifting a brick with a wax match.[18]

'Leaving the car,' one witness wrote after the quake, 'we walked along the Parade past five hundred yards of burning buildings, right up to the band rotunda in front of the Masonic Hotel. But the roof of the rotunda was on the ground, and all that remained of the Masonic Hotel was some gaunt gapped walls round which flames were leaping'. This picture was taken a day or two later. (JOHNSON COLLECTION)

'The business area of the town was desolate beyond anything I had ever imagined,' one witness wrote in the *New Zealand Herald* about Napier, the morning after the quake. 'It really was an inferno. Smoke was rising everywhere. Tennyson Street, Emerson Street and Dixon Street (sic), making up three sides of the principal block, were impassable.' Many witnesses compared Napier with the ruined French villages they had seen on the Western Front a decade and a half earlier. Sydney Smith took this picture after the worst of the debris had been cleared.
(Sydney Charles Smith, S.C. Smith Collection, Alexander Turnbull Library, G-48343-1/2)a

Police and volunteers search through debris after the quake.
(Brent Long, Brent Long Collection, Alexander Turnbull Library, PAColl-5330, F-20052-1/2)

ABOVE: Bluejackets at the bottom of Shakespeare Road. Tractors were common; other transport included an old Ford truck, dug out of the rubble, cleaned up and nicknamed 'Wheezing Winnie'.
(PHOTOGRAPHER UNKNOWN, P.T.W. ASHCROFT COLLECTION, ALEXANDER TURNBULL LIBRARY, MS-PAPERS-2418-5-02, F-139893-1/2)

RIGHT: Quake debris in Napier.
(PHOTOGRAPHER UNKNOWN, ALEXANDER TURNBULL LIBRARY, F-17217-1/4)

The Provincial Hotel was a 'heap of smoking bricks', where searching police found a 'calcined and crumbling bone' — which, with a pile of melted coins, was all that remained of the victim.[19]

Local officials and councillors held an initial meeting on the afternoon of 3 February, but serious decisions were not taken until early next morning when, with smoke still rising from the town centre, local authorities met Cabinet representatives in the Council Chambers. Parliamentarians who hastened to the 7.30 a.m. meeting included local member W.E. Barnard, Minister of Lands E.A. Ransom, Minister of Health A.J. Stalworthy and Opposition members, including Walter Nash, who had family in Napier, Peter Fraser and Robert Semple. Local authorities were represented by Napier Mayor J. Vigor Brown, Police Commissioner W.G. Wohlman and Hawke's Bay Hospital Board Chairman C.O. Morse, among others.

This group established a Napier Citizens Control Committee. Wohlman was appointed chairman, backed by Barnard, W. Harvey and J.C. Bryant, though he soon relinquished the position to Morse. Other members included representatives of the Defence Department, Police and Salvation Army. They faced a huge task. As Morse later wrote:

> … practically all public services had broken down, and a large number of persons had been killed and injured. It was necessary, firstly, to succour the injured, to feed the public, and to evacuate as quickly as possible the majority of women and children, both in order to allow them to recuperate away from the scenes of disaster, and also in order to prevent the out-break of disease …[20]

Sewage was a particular worry. The borough engineer advised that in the absence of power, neither the centrifugal pumps nor the pneumatic ejectors were operable. Even if they were, pipes on the flat had been smashed, and those on the hill were 'cracked and in parts badly broken'.[21] Borough water was equally problematic. The Cameron Road reservoir had been cracked, and several artesian wells were broken off below ground level.

Getting the health message out required help, and the committee turned to the local newspaper, the *Daily Telegraph*. They had lost plant and premises, but the Ball & Company printing office in Dalton Street was largely intact. Here a team of six *Telegraph* compositors set to work on 4 February to prepare a free broadsheet. This provided the committee with the vehicle it needed to relay the health message. Urgent notices in the first issue on 4 February warned 'that sanitary conveniences must not be used … The co-operation of the public is absolutely necessary.'[22] The compositors were interrupted when marines dyna mited the building next door, but were still able to publish the broadsheet. The building was condemned next day and production moved to Te Awa School.[23]

There were similar scenes in Hastings, where the Borough Council met again outside the damaged chambers. Henderson proposed a motion of sympathy for the relatives of those who had died, which was 'carried in silence, those present standing'. Councillor C.H. Slater was put in charge of the subcommittee dealing with food, based in King's Theatre. Attention turned to law and order, salvage and the need to get cars. Relief trains were also considered, though it was 5 February before a rake of trucks and 'eight empty carriages' reached Hastings.[24] A party of bluejackets arrived under Lieutenant-Commander Terry, and got to work searching the ruins for survivors, clearing debris and scouring the town for food.

Relief in Havelock North began on the day of the quake, from the town's own resources. This was not too surprising: socially, the special community bonds of what residents called 'our village' provided a stronger and more immediate source of succour than the general legacy of wartime experience that others in Hawke's Bay were able to draw on. Although its statutory authority was more limited than that of the neighbouring borough councils, the Havelock North Town Board acted at once, not waiting for official sanction and dipping into its own overdraft to pay for urgent repairs and clean-up. Food was made available by distribution from a central store. Board Chairman Herman von Dadelszen formed a local Relief Committee. Town Clerk W.B. Anderson, with Archibald Toop and E.F. Leicester, was instructed to inspect the water, power and sewerage connections.

EMERGENCY HOSPITALS AND MEDICAL AID

News of the disaster was quick to spread, and partly in response to Forbes' call for medical help, many medical people took their own cars or hitched lifts with others in the hope of getting to the disaster zone. Twenty men from the Auckland St John Ambulance station left for Napier on the evening of 3 February, with two cars and two ambulances laden with medical supplies.[25]

M.D. MacNab, a Wanganui nurse, had been 'night specialing' a patient in a private Wanganui hospital on the night of 2 February. As she was 'looking forward to spending the following night in bed' she 'did not go to bed that morning'. She felt the earthquake just before 11.00 a.m., but 'thought it purely a local shake and dismissed it from my mind'. That changed after lunch when she:

> ... went to town to do some shopping and, seeing a crowd at the windows of the local press office, stopped to see what it was all about. It was to say that disaster had struck Napier and many were dead and many injured, and nurses [and] doctors were urgently required. So I went in and gave my name. I was told that I would be contacted when transport was available. Early in the evening I was told that I was to travel with a Mr — who would be driving to Napier to find his two sons who were with his late wife ... I will never forget arriving at Dannevirke where the whole town was ablaze with lights and in a state of great activity. So far we had seen no great signs of damage beyond cracks in the roads so we stopped to ask for news & were told that Napier was on fire, that there were many dead and more injured and that they were 'operating with Pocket Knives and Razor Blades', those being the exact words used by our informant![26]

MacNab arrived in Napier around 4.00 a.m. on 4 February and reported at once for duty. Conditions were considerably better than she had been led to believe:

> Canvas wards were being set up and the St John ambulance produced what equipment they could for us and they did a wonderful job. But improvisation was the order of the day and it was not unusual to see wounds soaking in Kerosine tins of anti-septics [sic] as nothing else was available. I have no recollection of eating and drinking that day, but no doubt we had something.

Another early arrival was Dr Agnes Bennett, who reached the Marine Parade to find it partially blocked with 'bricks, plaster … and other debris'. On one side were the skeletons of burned out houses, on the other the sea wall which had:

> … fallen out in great chunks in places and against it every conceivable article of household use in groups — usually their owners sitting with them, an expression of dumb misery … at intervals some were making fires and attempting some kind of cooking. Beyond the sea wall … [tents and tarpaulins] … had been rigged for shelter … even the children seemed toneless and wearied ...[with] grief. A … self constituted traffic manager guided us on to the footpath and with difficulty we threaded our way through pedestrians …[27]

THE NEW ZEALAND HERALD

EARTHQUAKE HAVOC.

THE KILLED AND INJURED.

SUSPENSE AND DOUBT.

NAMES OF THE IDENTIFIED.

EVACUATION ORDER ISSUED.

A general order has been issued demanding the evacuation of Napier within two days. The authorities fear an outbreak of serious disease if the inhabitants do not leave the town.

In common with thousands of other New Zealanders the majority of Auckland residents spent a day of anxiety, suspense and doubt yesterday. The earthquake area was isolated, due to the entire collapse of the telegraph and telephone services.

It was known that many had perished in Napier and Hastings, but fluctuating estimates of the number of killed and injured in each of these centres failed to encourage or allay the hopes or fears of relatives and friends of those residing within the stricken areas. An official Wellington message received at a late hour gave a considered estimate of the dead in Hastings as between 80 and 100. Twenty-one dead have been identified in Napier, and 48 in Hastings, where six bodies, not identified, have been recovered.

The cruisers Dunedin and Diomede, which left for Napier on Tuesday afternoon, arrived yesterday. Commodore Geoffrey Blake immediately took charge of the relief measures. The efforts that were being made to stop the spread of fire among the ruined buildings of Napier were indicated in messages which passed between Commodore Blake from the shore and his ship, the Dunedin. Commander Clover, of H.M.S. Philomel, at the naval base, Auckland, reported that H.M.S. Veronica had intercepted the following message from the Commodore to the Dunedin:—"Send the lieutenant in charge of the torpedo stores, with all available demolition stores, to report to the Veronica." Commander Clover said that that indicated that it was intended to blow up several buildings in order to check the flames, which were still raging.

A message was also received by Commander Clover from the Veronica last evening stating that the ship Taranaki had sailed from Napier for Auckland at 4.30 o'clock yesterday afternoon with refugees. The Northumberland sailed at the same time with refugees for Wellington. The Taranaki is expected to reach Auckland at 7.30 o'clock this evening and will berth at Queen's Wharf.

Many other medical personnel followed — so many, in fact, that some were sent back.[28]

Conditions in the emergency hospitals improved during 4 February. MacNab worked all day, despite not having slept for two nights, and as evening drew on found that 'a row of sacking privies' had been set up. People stood in queues 'irrespective of sex' to await their turn. Later in the evening a marquee was erected for the nurses to sleep in, with groundsheets and blankets. MacNab and two others were the last off duty and found only one blanket left for the three of them. With that and a groundsheet each they found their way to the marquee, which was:

> … full of sleeping bodies. We dossed down together under the one blanket and cuddled close as although only February, it was a very cold night. Before lying down as the others did, fully clothed, I said I was going to undress as I had not had my clothes off for two days so I stripped in the darkness and donned my pink silk pyjamas. This was something I was afterwards to regret! Meantime, however, we were soon asleep and all too soon morning

dawned and we were called to report for duty. In that early morning light we sat up, surveyed our surroundings and our many bed mates. It was a very large marquee with a centre pole. At the foot of this centre pole a large figure erupted from under a blanket — a huge male figure with enormous walrus moustachios! The only non female figure in that concourse! He had been snuggled up to on all sides in the darkness by uncomprehending innocent girls of all ages, sizes and shapes! The nearest to him, a wispy little person with her hair in little piglets turned on him like a virago for ... endangering her reputation let alone her virginity when he insisted that, in the chill of the night, she had spent as much of it in his bed as in her own!

The man was the cook from the *Diomede*, and had arrived when the tent was empty. MacNab now found her pink pyjamas a problem: there was nowhere private to change. Eventually, after wandering around the racecourse, she 'took refuge in a horse box which proved very satisfactory'.[29] Next day the nurses were given 'hospital type iron beds and blankets' in the totalisator office. However:

... the earth tremors were so frequent and so violent that our beds rolled to and fro on the floor and sleep was impossible, so three of us decided to pull our beds outside under the trees. This we did, and settled down, but not for long — there were birds resting in the trees above and we received some unwelcome contributions on our uncovered heads. As a last desperate effort we again rose and pulled our beds out into the open. Before falling asleep I remember hearing my companion sighing with relief as she tucked the rather stiff grey blanket around her neck. In the morning light she found it was stiff with congealed blood.[30]

Some of the wounded were taken initially to private homes, including Len Greenfield, who had received serious chest injuries during the quake and was taken to a house in Napier's Nelson Crescent. He was picked up by a Palmerston North private surgeon and his head nurse, who drove to Napier to see what they could do but were deemed surplus to requirements at the emergency hospitals. They collected Greenfield instead, taking him back to Palmerston North where he made a full recovery.[31]

Medical people were not the only ones to arrive. The *New Zealand Herald* sent a 'special reporter' to the disaster zone on the day of the quake. He found the main road passable as far as Te Pohue but then had to take the Seafield road via Rissington. He reached Napier at night to find it 'a city of the dead, except for the glow of a land fire, and the lights of ships'. People he met had 'eyes ... bloodshot with sleeplessness'.[32]

EMERGENCY RELIEF — FOOD AND HOUSING

By the morning after the quake many Napier townsfolk were sheltering in tents on the Marine Parade beach and in other open areas around the town. Borough engineer W.D. Corbett organised water for them in barrels filled from 400-gallon (1820-litre) tanks on lorries, themselves topped up from broken but still serviceable wells in McLean Park. A visiting doctor watched this water arrive on the Marine Parade:

> People come running over from the beach to fill cups, jugs, bedroom ewers and buckets. A brewery has been raided and all the vats and casks taken out and distributed on stands at intervals. The water drawn from them deposits a sediment of what looks like dry hop leaves ... There is a distinct brewery flavour. The filling carts have hard work to keep these casks full. It is a thirsty day.[33]

During the morning of 4 February the Citizens Committee authorised several food depots: one at the Carlyle Street school under Ivy Maynard and her brother; another in the Hastings Street school, initially run by the *Northumberland*'s crew; a third at Miss Beharrell's school on Bluff Hill; another on Napier Terrace; and the last at Ahuriri. Naval ratings scoured the town for supplies, and Salvation Army Brigadier Greene was put in charge of distribution. These depots continued to operate until 20 February.[34] Kitchens were set up at the Hastings Street and Ahuriri centres. Another opened at Shortlands on the Marine Parade, to serve government officials and official visitors. Agnes Bennett visited the Hastings Street school soon after the kitchen had been set up:

> [I found a] scene of great activity — a soup kitchen was being run apparently by the sailors off one of the ships and one wondered if they were from the *Taranaki* or *Northumberland* — in rows of desks refugees were being given vessels of soup while inside in a classroom bread was handed out in 1/2 and 1/4 loaves, eagerly carried off in the hand ... notices were up 'please don't ask for more than you need.' Later we found that bakers in Dannevirke, Woodville and other towns had been up all night baking.[35]

These arrangements had been made by Minister of Health A.J. Stalworthy, who visited Waipukurau Mayor Robert McLean on the night of 3 February while travelling to the disaster area, and 'discussed vital matters of the moment'. He also organised a casualty clearing station in Waipukurau. More than 500

Outdoor cooking at the family home of Napier bookseller W.E. Storkey, Faraday Street. F.C. Wright peels potatoes; Margaret Storkey tends the fire. Collapsed kitchen chimneys forced most survivors to improvise cooking facilities in the days and weeks after the quake. Some fell back on food depots and communal kitchens for their meals.
(STORKEY FAMILY COLLECTION)

View from the Storkey home across Faraday Street: the remains of one of the few houses to actually fall during the quake intrudes at the top of the frame.
(STORKEY FAMILY COLLECTION)

Quake survivors with their possessions in the grounds of Bishopcourt, Napier.
(PHOTOGRAPHER UNKNOWN, WILLIAMS FAMILY COLLECTION, ALEXANDER TURNBULL LIBRARY, F-29571-1/2)

Wellington photographer Sydney Smith took this picture of an impromptu refugee camp on the railway side of Wellesley Road. There is little sign of the devastation in this view aside from the twisted chimneys. In fact the brick-built gasworks visible above the tents in the centre of the picture was badly damaged.
(S.C. Smith Collection, Alexander Turnbull Library G-48311-1/2)

Another row of tents popped up on the Tutaekuri river side of Georges Drive, then the edge of Napier South and the end of the residential district, where they were photographed by Henry Whitehead (1870–1965). The limed road is notable – dust was kept down in summer with a truck-towed sprinkler. The photographer was standing not far from the railway line, looking southeast down the road. Napier's premier 'moderne' housing district was later erected down Saunders Avenue, just to the right of this picture.
(H.N. Whitehead Collection, PAColl-3068, Alexander Turnbull Library, G-21339-1/4)

patients were evacuated through the station in the next days, and some of the more seriously injured went to the Waipukurau hospital.[36]

A Central Relief Depot was set up in the Hastings Drill Hall on 4 February. It did not take long for the system to be exploited, as E.F. Scott reported:

> ... it was found that a few were prepared to take all they could get and ask for more. On the second day, therefore, the Food Controller decided to make only one depot available for free food supplies, to stop the particular few from 'going the rounds'. And if you stop to think for a moment, there was no need for *everyone* to receive *free* stores. Quite a large number of people as for example, members of the Public Service, Municipal Employees, and others whose wages and incomes were not affected by the earthquake, were in a position to pay for their requirements ... The banks, however, had all closed down ... and the Citizens Committee had to devise ways of meeting the situation.[37]

On the Friday following the quake the Food Controller recommended that all food in Hastings should be commandeered and distributed by the committee; voluntary workers would get free food rations in return for labour; and the

In Havelock North survivors lived in their own back gardens for days. Apiarist Bill Ashcroft and his father built this 'dining room in our tent' outside their twisted Havelock North home.
(P.T.W. Ashcroft Collection, Alexander Turnbull Library F-139899-1/2)

Emergency arrangements in Hastings were very similar to those in Napier, though there was friction between Hastings Borough Council efforts to provide food via a single Food Controller, and the Red Cross. Initial Red Cross notices advertised 'Food galore, clothes galore, ask for everything you want.' This was well-intentioned, but the organisers soon had to borrow from official depots to meet demand. This depot was visited by Wellington photographer S.C. Smith a few days after the quake.
(S.C. Smith Collection, PAColl-3082, Alexander Turnbull Library G-48308-1/2)

The Hastings relief effort was centred at the Drill Hall and nearby buildings, where the Borough Council met several times before relinquishing the premises in favour of the Wesley Hall. War experience provided an invaluable organisational resource in Hastings. Local boys high school principal and former First World War officer W.A.G. Penlington ran the pickets, Colonel A.J. Manson organised transport and petrol, Colonel R.L. McGaffin distributed tents, and Colonel C.H. Slater, produce merchant, was in charge of the food. Major Justin Power was in charge of the Red Cross depot.
(S.C. Smith Collection, PAColl-3082, Alexander Turnbull Library G-47685-1/2)

sick could obtain certificates for free supplies. Food was commandeered from local shops. Ration cards were issued, and restaurants were set up offering cheap meals.[38] All the stock was accounted for later and unused supplies were returned to their original owners. £4600 ($331,200) worth of goods were obtained from one store, though about £1000 ($72,000) worth could not be accounted for and 'it had to be assumed that £1000 [$72,000] was the value of goods stolen or destroyed'.[39]

Clothes were another priority. Many survivors had only what they stood up in — ranging from pyjamas to street clothes and even formal wedding attire — and the executive committee set up a clothing depot in the Hastings Street Choral Hall, staffed by the Salvation Army.

POST, POWER AND RAIL

New Zealanders from Kaitaia to Bluff were clamouring for news of friends and relatives in Hawke's Bay. In the region itself, thousands of survivors wanted to let relatives know they were alive or convey the sad news of deaths. An officer was appointed to deal with queries in Napier, though his work was made 'extremely difficult on account of the people having left their homes'.[40]

THE NEW ZEALAND HERALD,

NAPIER LIKE YPRES.

TRAGIC QUEST FOR DEAD.

MANY MORE IDENTIFIED.

REMOVAL OF THE REFUGEES.

TERRIBLE STORIES NARRATED.

The earthquake toll is mounting, and the bodies of many victims were discovered yesterday.

Most of these have been identified. The number of dead amid the ruins is not yet known. Many have been injured; numbers are still missing. Early this morning nearly 200 bodies had been buried or awaited burial. Napier is described as resembling Ypres.

The accounts to hand give many instances of stark tragedy and amazing fortitude and heroism. Dr. A. D. S. Whyte, of Hastings, for instance, carried out urgent operations, although knowing his daughter lay dead amid the ruins. The injured bore their sufferings without complaint. The bluejackets of the Navy and hundreds of civilians carried out a tragic task—the quest for living and the finding of the dead—almost without respite. Further earthquakes punctuated the gruesome hours. A milkman attended his rounds at the usual time, but his milk—so much needed—was given free. He too must take his place in the front ranks of the gallant band of helpers.

Following the evacuation order, a great exodus from Napier has begun, and the scenes on the traffic-packed main road leading from the stricken town are indescribable.

Estimates of the damage, which is colossal, will be made later. One forecast places it at £2,500,000 in Napier, and £1,000,000 in Hastings. The Greenmeadows seminary has suffered to the extent of £50,000.

One hundred children of school age, from the earthquake area, are being sent to Auckland, and will be cared for at the Sunshine Association's camp at Motuihi Island. The children will be accompanied by 11 teachers. Provision for refugees is being made in many centres. Twenty arrived in Auckland from Napier by the steamer Taranaki last evening.

Destruction of the postal centres did not help. The new post office had been gutted and the Ahuriri post office had collapsed and burned. At first the Napier staff moved to the band rotunda in Clive Square, but this was commandeered as a surgical station, and postal work had to be split between the railway station and the Hastings Street school. Cars left with mail for Auckland on 4 February, reaching Taupo via the Rissington route by 2.45 p.m.[41] Other mail was despatched south by truck in the early afternoon of 4 February. Napier finally got into telegraph contact with Wellington around 1.30 p.m,[42] and a telegraph station was set up in the Hastings Street school. Bennett arrived soon afterwards:

> In another room of the school so miraculously saved from quake and fire ... A queue was formed and a harassed looking staff was doing its best to cope with the numbers ... the scene was like a very busy time in an election booth. There and in post offices as far south as Waipukurau one was impressed by the tired harassed faces of PO officials who apparently were working far into the night.[43]

Telegraph lines to Auckland via Taupo were brought back into action during 5 February.[44]

In Hastings the mail already in the post office was moved to the back of the building before being shifted, at around 11.00 p.m. on the day of the quake, to 'the departmental lineman's shed in St Aubyn Street'.[45] Other mail was in transit when the quake struck — the train carrying it stopped at Takapau. Until the line was inspected no trains could run, and the mail was taken by truck to Ormondville where a train came up from the south to take it. Little or no mail seems to have been lost during the quake.

Power was restored to Napier late on 4 February. Borough engineers were hampered by the loss of their store, burned out that morning, but managed to repair three of the seven single-phase transformers in the borough Power House, refilling the cases with oil and standing them on the ground. These were operable by 5.00 p.m., awaiting only the restoration of the grid supply from Redclyffe. This was not long in coming. Public works electrical engineers spent the day rigging cables to transmit 11,000-volt power from the 3500-kVA-capacity Waipukurau substation, down the 110-kV line and directly into the 11,000-volt Hawke's Bay regional ring main. All the connections were completed by early morning on 4 February, but it took the rest of the day to inspect the lines for damage, and power was restored about the same time as the Napier borough step-down transformers came back into action. Power was prioritised, starting with Napier — where it was urgently needed to get the water pumps working — then hospitals, street lighting and essential industries such as the freezing works, where export carcases had been thawing in freezers since the day of the quake.[46]

The railway was restored within two days. The shock had done a great deal of damage to the line north of Kopua Viaduct, about 50 miles (80 km) southwest of Napier. Rails had been bent and twisted and embankments had crumbled. Bridge approaches had subsided south of Hastings, and some bridges were strained. There was less damage between Napier and Hastings. North of Napier the line had again been twisted and fractured by seismic movement, but priority lay with the line south. All the bridges were found passable, the worst of the broken rails were replaced, and the first train reached Hastings around 2.00 p.m on 5 February, leaving two-and-a-half hours later for Wellington with passengers and mail. Next day the line opened as far as Napier, simplifying relief work even though repairs were good only for low-speed advance down the line.[47] The first priority was the injured. The Hastings racecourse hospital was reduced to a dressing station and the patients were evacuated to Palmerston North.[48] This did little to relieve the workload of Sister Dillon at the mortuary — she collapsed of exhaustion on 11 February and was reported 'resting at racecourse hospital'.[49]

Burials in both Napier and Hastings began the day after the quake. Interments in Hastings were organised along military lines with the help of a Church of England clergyman who had been a chaplain during the war. Twenty-five bodies:

> ... were buried in a common grave the second day, but thereafter each was placed in a separate and numbered grave, with a small wooden cross giving such details as were known. Where no religious denomination was known, the surname was assumed to indicate nationality and the appropriate padre officiated.[50]

Some 101 Napier quake victims were buried with a simple interdenominational ceremony in a communal grave in Napier's Park Island cemetery. Fourteen were unidentified. In 1932 a memorial to those who died in the disaster was raised above the grave. 'Their sun has gone down while it was yet day.' (MATTHEW WRIGHT)

In Napier, the 60-year-old wooden courthouse on Marine Parade — twisted and lacking chimneys — was pressed into service as a temporary morgue. There was a general inquest on 4 February at which relatives were asked to identify bodies.[51] Next day a special ceremony was held at Park Island cemetery where the dead, including the unidentified, were buried in a common grave 'unless specially requested by the relatives'. Services were read by 'the ministers of the different religions',[52] led by the Reverend G.T. Robson of the shore base HMS *Philomel*.[53]

A STEADY CLEAN-UP

Many people compared the destruction in Napier and Hastings to what they had seen during the war in France. The favourite point of comparison in the media appeared to be Ypres, probably because it was well known to many New Zealanders.[54] Some even quantified the destruction in terms of their wartime experiences. 'Men who were at the war said that the damage done in 10 seconds was more complete than two or three months of shelling with big guns would have been,' Dorothy Campbell later wrote.[55] Another witness told reporters he had been 'right through the war', and that 'nothing that I saw in war time could have done so much damage over such a wide area in so short a time as happened at Napier.'[56]

Squads of bluejackets worked to demolish unsafe walls, usually with rope and a hefty pull. Police, firemen and civilians pitched in to help. Work was slow at first; nobody knew who might be trapped alive under the rubble. By 5 February hopes of finding survivors were fading. A Hastings patrol early that day heard 'groans which were clearly audible 100 yards [91 m] away' in a ruined building near the town centre. They made 'frantic efforts to locate the sounds', but a thorough search drew a blank and 'the victim had to be abandoned to his fate'.[57]

Ernest St Clair Haydon was one of many who helped clear the Napier town centre. On the Monday after the quake he was with his son Joe and a friend, the jeweller L.S. McClurg, looking through the debris of McClurg's shop to see what could be recovered. Two staff members had been plucked alive from the wreckage of the shop days earlier, and now McClurg hoped to retrieve some of his stock.

> I never saw a boy so excited in looking for hidden treasure and no doubt, we recovered several hundreds of pounds worth of diamond rings, etc., but I am afraid poor old Mack is quite a ruined man as most of his jewellery was not insured; not that insurance will be any good as unless you had a special clause in your policy you were not covered against earthquakes ... After working there all afternoon I seemed to get a suspicion that something was wrong and on pulling out a few bricks where I had been standing, I found the remains of another body.[58]

Most of those working in the town laboured through a haze of exhaustion. One day Haydon went to the Hastings Street telegraph office, where he 'could not remember what day of the week it was or the date'. He asked a friend who happened to be there, but 'he could not tell me either and so we had to enquire from the operator to find out that it was Wednesday, eight days after the earthquake'.

Haydon set up a small camp on his Puketapu farm, later writing that 'we finally secured enough tents and we had a very nice party of twenty nine in camp. One of the things that struck me rather strange about the earthquake was that for two days we never heard a bird cheep or sing. It was the second morning when they suddenly burst into song again.'[59]

Health dominated official thoughts in Napier. A notice in the *Daily Telegraph* news bulletin of 5 February outlined how to dig a temporary latrine. 'All cases of diarrhoea are to be reported at the Health Office, Police Station, Byron Street.'[60] Residents were exhorted to cover fresh food against flies and to 'secure advice on sanitary matters' to avoid spreading disease. Fish left stranded in their thousands in the drained lagoon were a particular worry, and special steps had to be taken to clear them up.[61] Threatening weather at the end of the week brought fears of rain and pestilence.[62] Doctors were assigned zones around the towns to inspect for sanitation problems. 'The public must be as clean and tidy as possible in their habits,' a public notice in another *Daily Telegraph* news bulletin admonished, 'otherwise we will have disease in Napier.'[63]

The problem was less acute further afield. The Director-General of Health came to Hastings on 5 February and organised a pamphlet warning of water contamination from sewage. Traces of petrol were found in the water supply, thought to have come from ruptured underground storage tanks, and fuel oil was discovered in the sewers. Three days after the quake a Medical Officer of Health arrived and 'suggested that chloride of lime be used with the water as a precautionary preventative measure'. A chlorinating plant was later set up in the Power House.[64] In the end the incidence of disease was negligible — partly because of low disease rates before the quake, partly because of the prompt precautions taken to prevent its spread.

Fire remained a serious risk, particularly in Napier's housing districts. Late on the night of 5 February fire broke out in a Seapoint Road house on Bluff Hill when a copper in which a fire had been lit fell apart. The blaze 'threatened the whole of Seapoint Road' and was put out by sailors who 'did splendid work', dropping a bank onto the blaze while the cruisers 'played their searchlights on to the scene'. The fire took an hour to extinguish. A fire the same night in a Kennedy Road house was put out by the brigade. 'It is most unwise to light fires inside houses,' the *Daily Telegraph* news bulletin warned, 'and any lit in gardens for cooking purposes must be extinguished before citizens go to sleep at night.'[65] Not everybody heeded the warning: one man rebuilt his chimney with loose bricks and lit a fire, refusing to put it out until he was ordered to do so by marines.

Wairoa suffered a heavy blow on the night of 5 February when fire broke out in the freezing works. The borough water mains had been damaged by the quake, and although the brigade was quickly on site, it could do nothing once the emergency supply in the freezing works had been exhausted. The 'awe inspiring sight' was watched by 'hundreds of people' and destroyed 50,000 carcases being prepared for loading on the *Northumberland*. The blaze was punctuated by massive explosions as the ammonia tanks burst. The engine room and freezing blocks were destroyed and the fire tore through the pelt department, but for some reason left the offices untouched. 'Twisted iron and ashes were piled up in a huge mass eight feet [2.4 m] high.'[66] Two hundred men of the Wairoa Co-Operative Meat Company were put out of work by the blaze, which, coming on the heels of the earthquake, threatened ruin for the district. The cause of the fire was put down to an electrical fault in C-store.

THE BEST AND THE WORST

The quake brought out the best in some survivors, but in others it brought out the worst. Looting erupted within hours. 'A ghoul was seen stealing a wristwatch off a dead man lying on the pavement,' a horrified Dorothy Campbell wrote to her aunt a few weeks later, 'while articles of clothing were removed from living, wounded people. Houses were robbed while the owners were doing rescue work, goods & money were stolen from shops.'[67]

This was a different phenomenon from run-of-the-mill property crime, which was rare for most of the twentieth century. The crime rate actually dropped during the depression, and despite the extreme poverty faced by many in Hawke's Bay it was normal to leave houses unlocked.[68] Crime did not become endemic until the last years of the century — the per-capita rate rose sharply in the late 1980s, and by 1994 was fifteen times that of 1930.[69] Looting, by contrast, was a well-known accompaniment to any war or calamity big enough to disrupt the usual rules of society. Hawke's Bay after the quake was no exception, and authorities were quick to anticipate trouble. Marines were openly armed. Hastings' town centre was under citizen patrol by the night of the quake. E.F. Scott recalled that the volunteers operated in two six-hour shifts:

> [They] took up positions in groups of three at a time, with two watches a day at each point. They were supplied with pick and axe handles as batons and wore armlets and belts. The need for some sort of uniform was badly felt, as there were several attempts at impersonation ... However, certificates were given to every man, which had to be shown on demand, while patrols under an officer went round visiting the Railway Station, Relief Depots, and seeing that all pickets were on duty and were relieved to time. Patrols also watched out for any pilfering, but were hampered at night by lack of torches. Adequate supplies did not arrive till Saturday.[70]

Passes took some time to organise and in the end picketers were identified only by armlets. 'Workmen were taken through by their foremen,' Scott wrote, 'posters were pasted to the windshields of all official cars, and a list was supplied to every picket.' Special pickets were set up at the mortuary, and after a few days the site was specially isolated. As Scott recalled, 'the special police on duty were instructed to knock anyone on the head if they attempted to pass without producing proper authority'.[71] Even journalists trying to investigate the ruins were stopped. 'You cannot pass this way — martial law is in force,' one reporter was told in Hastings. Like all good newsmen he persisted, discovering that 'A little parley ... was possible.'[72]

Hastings patrols were extended to Pakipaki, Fernhill, Maraekakaho and Havelock North on 6 February 'to prevent ... persons coming into the Borough' and to stop empty cars leaving without passengers.[73] Some of these citizen patrols were armed more lethally than others. Bill Ashcroft later recalled that 'Bill and Ossie Wellwood were on one night and Bill got a waddie studded with nails. So if any looters were caught they were probably buried right off.'[74]

In Napier the Executive Committee published a notice in the *Daily Telegraph* news bulletin of 6 February warning that both looting and 'illegal trespass' were 'absolutely forbidden. The naval and marine guards are armed and will shoot on sight where looting occurs.'[75] Stories circulated that three men had been shot for looting,[76] but this did not stop schemes to relieve victims of property. 'Some people from other parts came in and tried to loot houses,' Bill Ashcroft wrote to his sister. 'Their idea was to set fire to a house and loot others while the owners were watching the fire.'[77]

The same day a call went out in Napier for 'able bodied men' to report to the police station to form patrols.[78] They were in operation alongside the marines by 7 February, 'picked men, who will possess the power to arrest and detain any person suspected of criminal intention'.[79] Honest citizens may have grumbled but were reminded by the *Herald Tribune* that 'pickets and patrols are protecting your life and property ... Help them and don't defy them.'[80] For many the pickets were a reassuring reminder that law, order and authority had survived the disaster. One Havelock North resident recalled the 'comforting sound of the night patrol, as their running feet pounded round the house every night'.[81]

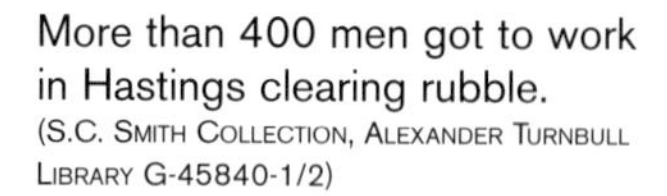

More than 400 men got to work in Hastings clearing rubble.
(S.C. Smith Collection, Alexander Turnbull Library G-45840-1/2)

Sightseers became a problem, reaching the point where by 6 February road blocks were set up on the bridge near Ashhurst 'to ensure that no tourists, sightseers or other unnecessary traffic will enter the stricken area'.[82] The fact that criminals might be among the rubberneckers was not lost on the authorities. A few days after the earthquake, police held an identity parade in Napier's Nelson Park. About 80 known criminals, who had arrived on the pretext of volunteering to help, were rounded up and sent out of town.[83]

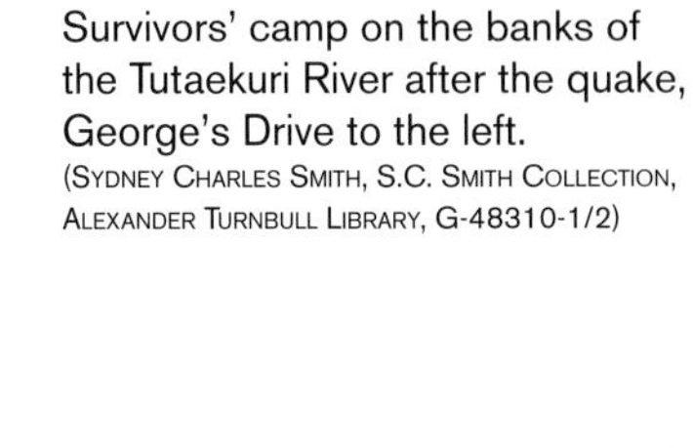

Survivors' camp on the banks of the Tutaekuri River after the quake, George's Drive to the left.
(Sydney Charles Smith, S.C. Smith Collection, Alexander Turnbull Library, G-48310-1/2)

Napier from the air some days after the quake, Nelson Park and the refugee camp centre of the frame.
(Photographer unknown, M.W. Buckley Collection, Alexander Turnbull Library, F-56953-1/2)

There were fresh problems in Hastings a few days later when stories circulated that men masquerading as civilian patrols were pilfering the ruins. On 9 February a concerned Mayor G.F. Roach asked Police Commissioner W.G. Wohlman for 'a force of 60 police officers to take over the guardianship of the town'. Wohlman demurred. Ten additional policemen had been brought into the region, of whom six were in Hastings, and 25 additional special police had been sworn in. He thought this was enough, though he was prepared to swear in a further 25 if necessary. 'I cannot see the necessity for a large force of uniformed men in this town,' he told the gathering. 'My information is that no crime exists … Exaggerated stories are in circulation.'[84] After a wrangle, police finally did take over from the civilians.

Thefts occurred despite these precautions. Napier chemist R.S. Munro returned to his wrecked shop to find that about £60 ($4320) worth of silver had been taken from the safe.[85] P.W. Barlow was very worried about theft from his home. He stayed in the Nelson Park camp for three days and made his way to his house twice every day to feed the chickens, 'haunted by the fear that some dishonest men from outside Napier might come with trucks and loot all the vacant houses in the street'. In the end he made his garage into a kitchen and built an outside fireplace so he could stay on his property.[86] But for all the opportunity the disaster offered the unscrupulous, most houses were left untouched: when Mary Hunter returned to Napier after nine weeks to start cleaning up her damaged flat, she even found the remains of some chops still in her meat safe.[87]

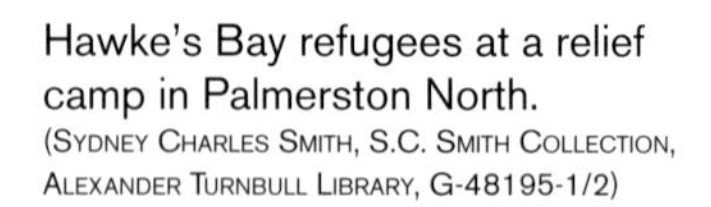

Hawke's Bay refugees at a relief camp in Palmerston North.
(SYDNEY CHARLES SMITH, S.C. SMITH COLLECTION, ALEXANDER TURNBULL LIBRARY, G-48195-1/2)

EXODUS, 4–7 FEBRUARY

A huge camp in Napier's Nelson Park provided emergency accommodation and an evacuation centre. It was organised by the Army early on 4 February using the equipment brought from Trentham on the night of the quake. The kitchen was operational by mid-morning. A refreshment room in the pavilion provided a steady supply of hot tea and was later expanded into a kitchen able to feed up to 800 people.[88] Red Cross workers and other volunteers arrived to help. Water squeezed from the aquifer by the quake was drained, tents were erected, latrines were dug, and rudimentary mess blocks were built to cope with up to 2500 people. Even water pipes were laid from a nearby artesian well.

At least 2000 people spent the night of 4 February in the park, and others arrived the next day. Although many were frightened, hungry and stressed, discipline never broke down and there was a strong community feeling.[89] Agnes Bennett thought the camp was a great morale booster:

> It was a pleasure to see the tents being set up with military precision — marines and bluejackets were in evidence and a good fire and a big oven gave promise ... Volunteer workers were busy and the Nelson Park camp was the most promising bit of organising that one had seen. The dull expressionless faces were disappearing and life and interest had begun to return.[90]

Entertainment was as critical as food and water. Two relief workers brought a marquee back from Wellington on 9 February. It was set up in Nelson Park and filled with seats from the wrecked Army Hall. A 'most efficient' orchestra and residents provided a 'galaxy of talent' for nightly shows, 'the psychological value of which was inestimably great'. By the following week free movies were being shown in McLean Park.[91] Thompson-Payne donated the projector, and the films were supplied free by Associated Film Exchanges.[92]

Few stayed long in the camp, which was mainly an evacuation centre. The decision to evacuate women and children from Napier was one of the first taken by the Napier Earthquake Executive Committee, mainly for health reasons. Palmerston North Mayor A.J. Graham had been at the 7.30 a.m. meeting in Napier on 4 February and offered to take up to 5000 refugees, with the help of vehicles offered by Manawatu Automobile Association members. Captain K.M. Findlayson organised the evacuation, later praised by L.G. Grant as 'undoubtedly one of the finest pieces of organisation ever carried out in this Dominion'.[93]

A *New Zealand Herald* reporter arrived while the evacuation was in full swing. 'It was a scene of infinite pathos,' he wrote, 'but also one which showed the spirit of service in full flower.'

> Vehicles were lined up from many places to convey the hapless people to succour and safety. Masterton, Dannevirke and Waipukurau were prominent names on the cars ... The crowd sat patiently on the grass surrounded by their poor, few parcels of luggage, packed as often as not in a sack. Here a mother nursed a little baby; there a girl limped along, a blood-stained bandage on her head ... A boy trudged in with an arm in a sling and an old bag in his hand ... An old man sat on a box and looked into oblivion ... A bright-faced woman met a relation from a distance and told him that her family fortunes could be bought for sixpence ... The marshal shouts 'Those for Dannevirke' and a group comes forward to be packed into limousines, lorries or buses ... Perambulators are got on board, fathers kiss their wives and bairns, and with brave smiles, not always hiding tears, they move away ... It is difficult for eyes, even those hardened to harsh scenes, to remain quite dry.[94]

By the end of the first day a thousand women and children had been moved, and next day it was announced that all who wished to go would be taken. Some 4783 had gone by 7 February. Not all went to Palmerston North — some ended up in Wanganui and other places, many staying with relatives or friends. Dorothy Campbell's family left on 5 February, driving through to Feilding with 'Ann, nurse & myself & Ann's cousins Hugh & his nurse'.[95]

Some found the exodus reminiscent of war-torn France. 'The traffic on the road to Palmerston North was tremendous,' one reporter wrote. 'Cars, lorries, everything on wheels seemed to be there. They were full of furniture, bedding, mattresses, babies' prams, and all sorts of household utensils. It was a flight.'[96] Mary Hunter was driven south by a Feilding dairy factory manager at night, in a convoy of 'thousands of cars almost touching each other', though she saw 'almost as many going to Napier'. The lights of the vehicles 'looked like a fiery serpent when we got to the bends in the road'.[97] Hunter's estimate of numbers was no exaggeration. The 70,000-odd residents of Hawke's Bay owned just over 10,500 vehicles in 1931,[98] but literally tens of thousands of other vehicles drove in during the two days after the disaster to bring aid or evacuate refugees. Some 27,000 cars passed through Waipukurau in thirteen-and-a-half hours on 5 February alone, marshalled by G.T. Clarke.[99] These numbers reveal the true national scope of the response.

Some refugees were billeted in a Waipukurau refugee camp. Others went to Palmerston North. A few made their way south independently in their own cars, including a Napier family who:

> On the day following the disaster all packed themselves into the car in which the kiddies had slept and drove to Dannevirke, where they asked the first man

> they met if he could tell them where they could get a tent to camp in the domain. That man insisted on taking them to his home, and kept them there for days under the assurance that if they left he would find more refugees to fill the house.[100]

Still others went to their relatives, among them Mary Hunter who stayed with her sister for the next nine weeks.[101] Evacuation came as a surprise for some. George Brown, who had helped medical staff on Napier Terrace on the day of the quake, was woken by people chivvying his family from the house. He went to the Nelson Park camp and soon afterwards took his family to Waipawa. He had not slept for several days, bathed for three or shaved for four.[102]

A few were evacuated by sea. Rona Lawrence was taken to the *Northumberland* but later transferred to the *Ruapehu*. 'It was an unusual experience,' she later wrote, 'sitting in a sling and being lowered over the side of the vessel to smaller boats which took us across to the *Ruapehu*.'[103] Some children were sent to the Sunshine Camp on Motuihe Island near Auckland.[104]

Just under a quarter of the refugees went by train — 1156 people were railed south after services resumed on 5 February.[105] Ernest St Clair Haydon was very admiring of the 'wonderful organisation' throughout his journey to Christchurch. At Waipukurau he and his family received donated cakes and tea. Red Cross nurses visited the train in Palmerston North, and the refugees got the best saloon accommodation on the ship south. He did not stay long in Christchurch — he was worried about his farm:

> Everybody's stock is all mixed up and as I have over two miles of frontage on the inner harbour which has now risen some seven or fourteen feet and the majority of it is now dry, it means that my stock can just stroll anywhere. It will be a big problem to know how to fence and confine them to our own property.[106]

In Hastings the water and sewerage systems were in better order, and evacuation was not a priority. Holderness reported to the council that 'he had only two applications for transport from people wanting to get away … and he suggested that the people should be encouraged to stay here meantime'.[107] In the end a few people did leave Hastings, but the number that had gone by 12 February was 'not more than 300'.[108]

A few people also left Havelock North, though — as in Hastings — not many felt they needed to go. The devastation was considerably less than in Napier, and the special community spirit provided a firm basis for close cooperation. Village resident S.M.M. von Dadelszen later recalled that 'one of my warmest memories of the earthquake time is of the kindness and helpfulness of everyone ... The quake seemed to have brought us all closer together as a community, and all the artificial barriers between people were dissolved.'[109] People camped in tents or makeshift shelters in their gardens. Emmy Turner-Williams organised open-air performances of several Gilbert and Sullivan operas by way of diversion:

> Four lorries were backed together in the Presbyterian church grounds. The cast rehearsed as best they could, and the costumes were gathered together from many cupboards and fancy dress collections around the village. The music and songs were played by gramophone and amplifier and the cast acted and mimed the words. It was all great fun.[110]

Sightseers take a closer look at the slip below Napier's Bluff Hill.
(HENRY NORFORD WHITEHEAD, H.N. WHITEHEAD COLLECTION, ALEXANDER TURNBULL LIBRARY, G-21337-1/4)

Napier's fire brigade in their temporary post-quake premises, with one of the two Dennis fire engines.
(SYDNEY CHARLES SMITH, S.C. SMITH COLLECTION, ALEXANDER TURNBULL LIBRARY, G-48315-1/2)

Hopes were still held for missing people even days after the quake. One of the miracle survivors was 90-year-old James Collins, pulled from the wreckage of the Park Island Old People's Home after three days under rubble, apparently not much the worse for his ordeal — though he told reporters that he 'could have done with a gallon of beer'.[112]

For others the story did not end happily. A young husband on his honeymoon frantically searched for his wife, who had been shopping in Napier's town centre when the earthquake had struck, but she had apparently 'been felled by falling brickwork and fallen victim to the flames'.[113] Sheila Hindmarsh's husband 'followed up rumour after rumour' looking for his wife. He went to all the field hospitals before he found her body in the Courthouse mortuary.[114] A plea was published in the *Daily Telegraph*'s earthquake broadsheet for 'any information' about Herbert Dennett, aged 60, 'wearing double-breasted navy blue suit ... not seen since quake'.[115] Other people were only identified by public appeal. Details of eight unidentified bodies were published in the *Daily Telegraph*'s earthquake broadsheet on 7 February.[116]

This seems to be a letter of horrors but even so I have not told you the worst & won't, they are too ghastly.

DOROTHY CAMPBELL TO HER AUNT, 5 MARCH 1931[111]

By 5 February the death toll was more than a hundred and still rising as teams of bluejackets and local volunteers worked to clear the rubble-filled streets.[117] It was exacting work. The debris had to be carefully probed before being cautiously lifted clear. Many hoped relatives might be found alive, although the *New Zealand Herald* warned people not to raise their hopes, advising on 7 February that 'it is definite that during the last two days no one has been dug out from the debris alive'.[118]

There remained the vital task of accounting for all the missing. Ernest St Clair Haydon found a body near the ruins of McClurg's jewellers as late as 9 February. There were fears that the toll would never be fully established, and he later wrote despondently: 'I think the present death rate of 250 and 2,500 severe casualties is just about half the true estimates, and they are still finding bodies everywhere — there were 15 boxes of charred bones and they will never know who they were.'[119]

No certain toll could be given even as late as 1933. As F.R. Callaghan wrote that year, 'it was extremely difficult to arrive at a precise number owing to various factors which are obvious in a calamity of such a character'. However, he was of the opinion that 'the unrecorded deaths' would probably not 'exceed ten at most'.[120] The story that a car had been buried with its occupants under the rubble of Bluff Hill was considered so reliable that a week after the earthquake there were calls to preserve the debris as a 'natural grave'. This happened — a new road was built round the 100-foot (30-m) slip,[121] and the idea that the rubble held bodies was not scotched until the 1960s when the last of the debris was finally cleared.

Of the 256 officially recorded deaths, 161 occurred in Napier, 93 in Hastings and 2 in Wairoa. Efforts to find and identify casualties were outstanding: only 28 people were unidentified when buried. F.R. Callaghan conjectured that up to 60 percent of victims died 'on the footpaths where they were overwhelmed by falling masonry' or by entrapment 'while attempting to escape from the collapsing buildings'.[122] The unofficial toll was 258.

For many survivors the days after the earthquake were filled with sadness, loss and even despair. Everybody had lost something — income, jobs, homes or property. But by far the greatest loss was human. Nobody in the district was untouched. Many had lost close friends or relatives, some before their very eyes. Everybody knew someone who had been injured or killed.

The appalling human loss took its toll on the survivors, who were in some cases already traumatised by their experiences. W. Olphert thought that after the quake the 'people of Napier were wonderfully cheerful',[123] but this would seem to have been the proverbial 'brave face'. Agnes Bennett thought the groups she found on the Marine Parade the day after the quake were wrapped in 'an expression of dumb misery ... even the children seemed toneless and wearied ...[with] grief'.[124]

A few collapsed immediately after the quake into what E.F. Scott called 'an attack of excitement hysteria'.[125] This was wholly understandable. Most survivors had gone through intense personal trauma, without warning or preparation. One woman escaped from Roach's store in Hastings only to see two friends crushed by the falling building. For some hours she and several other survivors wandered 'aimlessly about the streets, which were piled up with wreckage. We were all dazed, and could only sit there and half realise what had happened. The most terrible feeling was that of helplessness.' She was later found by a friend who 'took her to a place of refuge'.[126]

These were not isolated incidents. Another survivor, trapped in the doorway of a tea-room in Napier's wrecked Masonic Hotel, was so traumatised as to be scarcely recognisable. 'She seemed to have been turned into a different person in that horrible catastrophe,' a witness recalled.[127] Reporters arriving in Hastings on 4 February were appalled to find a 'half-demented woman aimlessly wandering the wrecked streets and asking people indiscriminately if they had seen her husband or children'.[128]

Everybody reacted differently. 'Aunty was hysterical absolutely,' Jean Anderson wrote to a friend soon after the quake.[129] Nor were military personnel immune to the overwhelming depth of the human tragedy. Training and a dispassionate separation from the disaster could only do so much. Mary

'Palmerston North has offered to provide for 5000 women and children, and the evacuation is proceeding,' the *New Zealand Herald* reported on 6 February. Wellington photographer Sydney Smith visited the well-organised site a few days later.
(S.C. SMITH COLLECTION, PACOLL-3082, ALEXANDER TURNBULL LIBRARY G-48690-1/2)

Hunter was helped a few days after the quake by a marine who was 'really shaking with horror'.[130] One sailor recalled that 'many of us had tears in our eyes and lumps in our throats as we carried out our duties'.[131] Another told a survivor he had 'never seen such appalling sights'.[132]

For Dorothy Campbell fear did not come until she and her family were driven from Waimarama. She had been 'too dazed to feel anything before', but:

> ... driving for ten miles over a bad hilly road in the dark with earthquakes going on all the time & dodging an occasional fissure in the road, it was awful. Of course what made it worse was the fact that we had none of us had any food since practically the night before, but somehow, however much one tried, it was just impossible to eat ...[133]

Stress and aftershocks made the nights a restless vigil for most. 'All those first nights we lay waiting for the next shock,' S.M.M. von Dadelszen of Havelock North recalled, 'and watched by the face of the cold, uncaring, full moon, which made sleeping even more difficult.'[134] Dorothy Campbell found that even two nights after the quake sleep was hard to find. 'When I reached the Palmers it was pretty late so we all went to bed, at least we lay down with our clothes on but I don't think that any one in the house except Ann slept, the earthquakes were too bad.'[135]

Bernard Chambers — a regular but terse diary keeper since 1874 — waxed almost lyrical during the days after the quake, perhaps a sign of the strain he felt. 'In aftrn. to the funeral of R. Mac. at Hav. Cemetry, he was crushed in his car in the street outside of Roach's,' Chambers penned on 5 February. Two days later he went to the funeral of his relative Mary (May) McLean, killed in Napier, writing that 'her body had been recognised in the Morgue at Spit. Arch. [Archdeacon] McL [sic: Maclean] read the burial service.'[136] For Chambers, the loss of his home and the general strain took their toll: he suffered a stroke just over three months later and died.[137] He was not the only one to suffer. 'Everyone's nerves, of course, were very much on edge,' Ernest St Clair Haydon later wrote to a friend:

> Although you did not feel nervy yourself in a way, it still affects you strangely. Whenever I received a telegram or cable I found that I was roaring with laughter and at the same time crying like a baby — a pure sign of one's nerves.[138]

Dorothy Campbell thought the 'dreadfulness of the whole business cannot be exaggerated', writing to her aunt of the:

> ... awful deaths there had been, one man we know being literally roasted to death in a cellar, one could only hope that the people who were killed were killed quickly. One little boy was imprisoned for two days & when he was dug out said: 'Gee I'm hungry'... looked around, gave one scream & ran away. He was never found again. There are cases of women and children, even men who have been driven quite mental with fright, the doctors are having their worst trouble with these. They say that they are similar to very bad cases of shell shock ...[139]

Many felt relief that they and their families had survived without loss. Dr Agnes Bennett found a 'new sense of values' was developing during the days after the quake. She commiserated with one family who had lost their possessions:

> … the mother said 'oh we don't mind, we are all here' — so the father, chopping wood close to the tarpaulin, under which they were all camping called out 'want to buy a nice cottage? A really nice cottage? Only trouble is it's buried under the cliff.' … Another mother of four just out of hospital herself said: 'I don't mind — I'm so thankful I have them all.' She came cheerfully away in a car with four children and some blankets — camped in a motorists camp, no food, no change of clothes, no possessions of any kind and yet thankful and convinced that she was well off … Many will point out how fortunate are they that the time and weather was such as it was — so changes our sense of values according to the circumstances surrounding us![140]

The Dominion-wide 'day of mourning' on the Sunday after the quake was marked in Hawke's Bay by special public services attended by the Governor-General, Lord Bledisloe. It was meant to be a day of rest, though as one reporter noted, 'really no-one can rest, particularly after a severe shake which occurred at 1.45'.[141]

AFTERSHOCKS

Aftershocks kept the trauma alive. Haydon later recorded that the ground was quivering almost continuously. Many of the shocks were only detectable by instruments such as the Milne-Jaggar recorder set up by Christchurch scientists at the Hastings racecourse. The smaller shocks were punctuated by larger events. A major aftershock around 9.00 p.m. on 3 February compounded the earthquake damage. Another, on the following Friday, sent more land slipping away at Mohaka and blocked the river altogether at Willow Flat.[142]

The biggest aftershock came around 1.30 p.m. on 13 February. With a magnitude of 7.3 and a Modified Mercalli rating of VII across most of Hawke's Bay it was New Zealand's fourth strongest recorded quake, centred about 30 miles (50 km) east of the 3 February shake. Power failed three seconds before it was felt in Napier. People from Napier to Dannevirke ran for their lives as damaged buildings cracked and fell. In Napier, Dr Moore's hospital took on a more catastrophic lean — while its owner was inside rescuing personal effects — and there was another slip on Bluff Hill. A wharf shed at Ahuriri collapsed and there was other damage to the wool stores.[143] At Tutira the shock sent Barbara Guthrie-Smith rushing out of the house, where she was knocked off her feet. Her father followed her out, 'banged from side to side from wall to wall during my exit', though he felt 'the movement was not more violent than that of a ship rolling heavily'.[144]

In Hastings the shock rattled the Power House, 'crushing the ends of the timber cross-bracing, necessitating further shoring', but the engineers did not shut the engines down.[145] Corrugated iron fell 'from precarious heights' into Heretaunga Street. Bells in St Matthew's church in King Street rang as the tower swayed, and part of the church collapsed. Fortunately the tower itself — tilting from the 3 February quake — only took on a slightly greater lean. Other buildings apparently untouched by the earlier shake were also damaged, including a block of flats which had to be demolished. Workers gathered to discuss their experiences before returning to the task of rubble clearing. Curiously, the sole chimney in the middle of town to survive the 3 February quake — on the Bank of New South Wales building — remained standing on 13 February 'without a brick out of place'.[146]

Some inland parts of Hawke's Bay felt this aftershock more strongly than the 3 February quake. In Taupo, goods were thrown from shop shelves, but there was 'no damage of any moment'.[147] People rushed into the streets in Dannevirke and Masterton. In Wellington all but one of the clocks stopped in the Dominion Observatory, and ceiling lights in the *Evening Post* offices swayed 'more vigorously' than they had the week before.[148] The shock was registered in Sydney, where the Reverend Father O'Leary of the Riverview observatory put it down to 'crumbling of the ocean bed' east of Napier. In New Zealand, government seismologist C.E. Adams reassured reporters that it was actually an aftershock. '[The] earth is apparently settling down after a rupture and is easing the strain by minor breaks,' he explained.[149]

CHAPTER 5

'... *an indescribable chaos of debris*'[1]

The prognosis for Hawke's Bay — and indeed for New Zealand — did not seem good in the weeks after the quake. The news sent stocks plunging on the London exchange,[2] and the possibility that the ruined province might drag a depression-wracked nation down was not lost on the editors of the *New Zealand Financial Times*. A notice in the February issue called for a full insurance payout, funded by government loans of £10 million ($720 million), raised in London and to be repaid through increased fire insurance premiums.

> Hawke's Bay has always earned a large share of the national income and borne its fair share of the Dominion's obligations. Unless it can be restored at the earliest possible moment, New Zealand must face a heavier share of the Dominion's burdens on a reduced national income ... The stupendous task of ascertaining the nation's loss ... must be pushed ahead with all possible despatch ...[3]

Others thought the quake would be the last straw. 'Farming was in a very bad way before this came along,' Ernest St Clair Haydon wrote to friends soon afterwards, 'and it now seems to have put the finishing touch to it all.'[4] He was not the only one to make this observation. C.E. MacMillan realised that the quake had compounded the effects of drought and depression. 'So misfortunes have not come simply,' he wrote to his wife.[5] A despondent Dorothy Campbell saw little hope for the future:

> The newspapers use the old slogans 'Business as usual', 'A newer & better Napier will rise from the ashes of the old', but oh the pitifulness of it all. One goes to Hastings and sees a man with perhaps his wife & a little child standing by & gazing at a tumbled mass of ruins. If one stops & speaks to him, saying 'I am sorry, was that yours?' he will probably reply 'Yes, everything I had was there.' People are still too dazed to take it all in. At present they are being fed with Relief Money which is being spent as fast as it comes in, but that cannot go on forever & what is going to happen then? How are these people to get work when there are already so many unemployed in the country? How are they to get money to start again in their businesses? Many of them are old & everyone

knows how hard it is for the old to start again. It is a tremendous problem & one can only pray that the Government will see its way to a right solution & will try to sink party differences in the face of a national disaster.[6]

HAWKE'S BAY RETURNS TO LIFE

Although the future seemed gloomy, those left after the evacuation worked to clean up the debris and restore services as quickly as possible. HMS *Veronica* left shortly after noon on 10 February. The harbour was too shallow for her to manoeuvre, so she was towed out to the cheers of watching Napier townsfolk. Damage to her rudder meant she had to be steered by hand, and she was escorted to Auckland for dockyard inspection by HMS *Diomede*. HMS *Dunedin* followed the next day.[7]

Many Napier businesses opened their doors during the second week after the quake, among them Acetone Illuminating and Welding Company Ltd, which reopened its Owen Street premises on 11 February. H.C. Motors 'resumed operations in their former premises' with 'full stock, and full services'. Harvey, Fulton and Hill opened a temporary office in the Hawke's Bay Motor Company building on Dickens Street.[8] Next day Humphries Cash Groceries reopened its four shops. Banks were authorised to reopen on 16 February. Napier's trading banks, clubbing together under the name Associated Banks, hired Fletcher & Co. Ltd to build temporary joint premises on the corner of Dickens and Munro Streets. Business boomed for some. 'As lots of people lost all their jam the demand for honey is very brisk,' Havelock North apiarist Bill Ashcroft wrote to his sister twelve days after the quake, 'and I'm extracting as hard as I can go.'[9]

Palmerston North photographer J.H. Daroux was commissioned to take pictures of the damaged buildings in the earthquake zone for insurance purposes after the quake.This is the Dalgety & Co. building on the corner of Dickens and Dalton Streets. It survived both quake and fire and remained a well-known Napier landmark.
(J.H. DAROUX COLLECTION, PA1-F-145-01-1, ALEXANDER TURNBULL LIBRARY)

F.C. Wright stands amid the wreckage in Napier's Clive Square.
(Storkey Family Collection)

Quake debris in Napier.
(Johnson Collection)

Twentieth century architectural styles were prominent in Napier before the quake. Some of the early examples – such as the Napier Ford Garage, seen here – were destroyed by the disaster.
(Photographer unknown, Pickering Collection, Alexander Turnbull Library, F-56958-1/2)

This is the Hawke's Bay Farmers Co-Operative Association premises at 43–45 Dickens Street, after the quake. Bracing put in place to support the weakened structure after the disaster is evident.
(J.H. Daroux Collection, PA1-f-145-05-30, Alexander Turnbull Library)

Most of Napier's newer buildings, erected with reinforced concrete or steel-frame structures, tended to survive the quake with minimal damage. This is the 'prairie' style Soldiers Club designed by J.A. Louis Hay and completed in 1920. This building, and others of the 1920s, established a strong pre-quake tradition of innovative architecture and up-to-date styling in Napier, setting the pattern for post-quake reconstruction.
(J.H. Daroux Collection, PA1-f-145-07-38, Alexander Turnbull Library)

Dr W.D. Fitzgerald's house at 13 Marine Parade after the quake.
(J.H. Daroux Collection, PA1-f-145-13-77, Alexander Turnbull Library)

It looked touch and go for the municipal building in Hastings. 'Brickwork bad and looks dangerous' was the insurance verdict. In fact the building was repaired.
(J.H. Daroux Collection, PAColl-5465, Alexander Turnbull Library C-22945-1/2)

The Morris Sales and Service garage in Hastings was 'badly shaken & cracked', according to the insurer. 'Sure to be condemned.'
(J.H. Daroux Collection, PAColl-3034, Alexander Turnbull Library F-4808-1/2)

Ross, Dysart & McLean's motor garage after the quake.
(J.H. Daroux Collection, PA1-f-145-62-101, Alexander Turnbull Library)

Produce in storage at the time of the quake was exported during the next week, mostly by rail. A good deal had been destroyed; however, 10,000 bales in the Ahuriri Loan and Mercantile store valued at 'between £70,000 [$5,040,000] and £80,000 [$5,760,000]' was shipped by 9 February.[10] Frozen meat was examined and railed out. Although the freezers in the four district works had been disabled, the insulation held and at Tomoana the auxiliary steam plant was brought on line. Damage to the port made the longer term prognosis less buoyant. The *New Zealand Herald* reported gloomily on 9 February that 'there is not even road access to the quay where the lighters used to load'. Bulk cargo had to go south by rail with higher freight costs, a heavy blow in the face of depression pricing. 'It is doubtful whether it will pay ... [farmers] to send lower-class animals, such as old ewes, to the works this season,' the *Herald* opined.[11] Nor could all the works restart operations. Borthwick's Pakipaki plant was a near-total loss. The heavy machinery was sent to Masterton and the Hawke's Bay Farmers Meat Company agreed to let Borthwick's process carcases at Whakatu. This was still being done as late as 1939. Heavy damage to the North British Freezing Works at Ahuriri was never repaired, and its plant was dismantled in 1933.[12]

Napier's Earthquake Relief Committee began considering ways of returning people to their homes. Some had already moved back by 6 February, prompting stern warnings about the health risks. 'People must not use water unless it is boiled, no matter from what source it is obtained,' a public notice in the *Daily Telegraph*'s daily news bulletin thundered on 7 February.[13] Next day an even stronger announcement forbade people to light fires inside their houses. 'This practice is dangerous and must be stopped. The Brigade has authority to take the names of offenders, against whom action will be taken.' Residents were again exhorted to boil water. 'Sinks and W.C.'s must not be used.'[14] By 10 February water had been restored to parts of the town near the sea. However, almost all chimneys had fallen and sewage remained a problem. Plans were soon in hand to inspect each house.

In Hastings most of the reticulation remained intact. Leaks still needed plugging, but water was back on across most of the borough by the end of the week. Temporary repairs linked the town with its reservoir in the Havelock hills, though it took a fortnight to refill the reservoir.[15] Sewerage was another matter: householders found that 'a surprisingly large number of the closet pans were broken'. Afterwards, one plumber alone ordered 150.[16]

Official government and bank records were generally damaged or lost. The Hastings branch of the National Bank suffered ill fortune of biblical proportions, escaping only a plague or two of locusts. The building survived the quake, but any thought that it might reopen was quashed next day when fire ripped through the structure. This did not touch the basement where the records were kept, but that only preserved them a further 24 hours because the basement was then flooded. The sodden securities and ledgers were finally retrieved the following Monday, though even then the strongroom had to be blasted open because the keys were missing.[17]

The first outside bulletins were published on 13 February, leading with the news that a week earlier Malcolm Campbell had reached the unprecedented speed of 245.75 miles an hour (395.49 km/hr) at Daytona beach in his Napier-engined *Bluebird* car.[18] That day marked the end of the *Daily Telegraph* broadsheet. The company obtained a press from the *Pahiatua Herald* the week after the quake — 'a Wharfdale two-feeder, six linotypes, a Ludlow caster, and a goodly supply of type'. The press was set up in the Vulcan Foundry. The linotypes went into a building over the road, and compositors carried assembled printing forms across the road. The first attempt to use this press on 12 February ran into mechanical problems, but on 14 February the paper produced its first full post-quake issue — a four-pager.[19]

Most businessmen tried to reopen as soon as possible after the quake. Arthur Dillon's tobacconist and hairdresser recommenced business literally in the middle of the rubble. (Alexander Turnbull Library F-17216-1/4)

Recovery brought new demands for power. The lash-up that sent 11,000 volts down the 110-kV line from Waipukurau to the Hawke's Bay regional ring main was limited by the 3500-kVA capacity of the Waipukurau transformer station. A spare 1500-kVA transformer was brought up by road from Mangamaire to meet demand until the Taradale transformers could be repaired. Damaged bridges meant the 20-tonne unit could only be got to within a mile of the Redclyffe substation, but it was put down on the banks of the Tutaekuri and hooked directly to the 110-kV main. Engineers then rigged an 11,000-volt line from the transformer to the Redclyffe substation, running it through the windows to the bus-bars. It was an ingenious piece of engineering that got a good supply back into the regional ring main by 22 February. Engineers then repaired the step-down transformers, finding:

> ... the main windings and cores were undamaged, but all the bushings were broken; radiator fins damaged; oil piping broken and all oil lost. Spare bushings and sufficient oil were obtained from other substations. The transformers were successfully dried out, given an over-potential test, and reconnected into service on the 29th March.[20]

Repairs were also made to the line from Tuai. All the transmission towers had survived with the exception of No. 220, the foundations of which had been carried away by a landslide. A deviation line was in place by 31 May.[21]

This picture of Napier's town centre from the hill after the quake reveals the sheer extent of the devastation. There were some early calls for a completely fresh start elsewhere, though in the end this was not done. The undamaged Public Trust building is just visible on the right, a rare survivor.
(KITTY WOOD COLLECTION, PACOLL-1009-06, ALEXANDER TURNBULL LIBRARY)

DEPRESSION FUNDING

As the days wore on thoughts turned from immediate disaster relief to the wider problem of rebuilding the shattered province. Government help seemed essential, and after some lobbying, 50 MPs arrived at the beginning of March to see for themselves what had to be done. On 11 March, retired magistrate J.S. Barton and retired engineer L.B. Campbell of the Public Works Department were appointed Commissioners of Napier, temporarily replacing the city council. With the help of Wellington city engineer G. Hart they had the task of restoring a town where almost all services had been destroyed. Sewers and drains had been so completely wrecked in Napier South that the whole district had to be replumbed. Houses had to be repaired, inspected and certified before they could be reoccupied.

Statutory assistance came the following month in the form of the Hawke's Bay Earthquake Act, passed when the House reconvened. Although 'an enormous volume of evidence' was considered during the Select Committee stage, the Act was still prepared in some haste and proved a mixed blessing.[22] While politicians had been quick to organise humanitarian aid in the immediate wake of the disaster — in part a legacy of personal experience in the First World War — longer term relief was characterised by the ideals of the depression. In the early 1930s politicians were flirting with discredited settler-era economic dogma, seeing a hands-off approach as the only way to stop New Zealand sinking in the face of the world crisis.[23] The Act limited larger scale recovery assistance accordingly. This was not well received in Hawke's Bay, and resentment grew through 1932 and 1933 as the full implications became clear. The unprecedented political swing by traditionally conservative Hawke's Bay electorates towards the Labour party in the 1935 elections almost certainly had its origin in the state's response to the 1931 earthquake, though other factors also played a part.

The Act created several new bodies to deal with post-quake financial confusion, notably the Adjustment Court, so named because of its role in 'adjusting' pre-quake liabilities. The legislation also gave the crown preferential claim in case of bankruptcy or winding up 'within five years'. Local authorities were generally required to fund their own recovery, though the Act did permit them to raise loans without recourse to ratepayers — ordinarily these things went to a poll — and gave them additional powers to raise overdrafts. Mayoral elections were postponed by the same statute.[24] In the end the government provided £1,770,000 (which translates to approximately $127,440,000 in late-twentieth-century money), including £840,000 ($60,480,000) to local businesses, £250,000 ($18,000,000) to local bodies, £200,000 ($14,400,000) for the reconstruction of public buildings, £35,000 ($2,520,000) for railways, £84,000 ($6,048,000) for roads and bridges, and £72,000 ($5,184,000) for surveys and maps.[25] This was nowhere near enough. Napier Borough Council took out a government loan of £101,200 ($7,286,400), a public debenture of £30,000 ($2,160,000) and a further loan of £30,000 ($2,160,000) from the AMP Society. The cost of meeting the interest payments alone was crippling, and the situation seemed hopeless until the Labour government released the town from its state debt in 1938. This prompted screams of protest in Hastings.

Havelock North seemed to have been left out of the picture altogether, and the Town Board fought a full-scale bureaucratic war to get funding for the close-knit community it administered. Board chairman Herman von Dadelszen estimated it would cost just over £4000 ($288,000) to repair power, water and sewerage, quite apart from recouping the costs incurred in the days after the quake. Government did not help much. 'Understand your department erecting one chimney each house Hastings free of cost to owner,' Town Clerk W.B. Anderson telegraphed the Minister of Public Works on 20 February; 'is this privilege to be extended to Havelock North?'[26] It took a fortnight to get confirmation, and when it came the offer was a question — did Havelock North want the same treatment as Waipawa? Von Dadelszen confirmed that it did, and asked for £100 ($7200) to cover post-quake clean-up costs.[27]

When this application was turned down, von Dadelszen launched into a spirited correspondence with Prime Minister George Forbes, who argued that the board did not deserve the refund because it had spent money on tasks other than street clearing. He was prepared to offer £50 ($3600) by way of compromise. Von Dadelszen persisted, arguing that in every other town the Public Works Department had cleared the streets 'without cost to the respective Councils, whilst here my Board did the whole of the work on its own initiative'. The battle went on for two months before Forbes finally turned him down flat. There was more bad news later in the year when the board tried to get money under the provisions of the Act. Only half the loan they requested was approved. The hard-pressed board had to scrape up funds from the State Advances Department and by issuing nineteen £50 ($3600) debentures.[28]

Hawke's Bay property owners were insured to the tune of some £5,000,000 ($360,000,000), which comfortably exceeded one loss estimate of £2,500,000 ($180,000,000). However, few policies covered earthquake damage and only about £250,000 ($18,000,000) was actually paid out by insurance companies.[29] Assessments led to some surprising conclusions. An engineer's report on Dr Moore's tilting hospital showed that it was 'worthwhile' to repair. Instead it was demolished. The Napier Public Trust building, by contrast, was written off as a total loss — but in fact was repaired.[30]

Most of the recovery money came from charity, which flooded in during the weeks after the quake. Auckland City Council had a fund started as early as 4 February, kicking it off with a £250 ($18,000) donation.[31] By the next day the Timaru Borough Council had offered £500 ($36,000), the *Christchurch Press* had started its own fund with £250 ($18,000), and the Nelson City Council sent £100 ($7200) each to Napier and Hastings by way of a 'first instalment'. Nelson's 'K' factory donated 6000 tins of jam and 3000 tins of tomato soup.[32] King George V and Queen Mary offered £750 ($54,000) each from their personal funds.[33] Sailors met a request for donations with generous enthusiasm. Some £530 ($38,160) was collected from the Navy — equating to 10/- ($36) per man, more than a full day's pay for many.[34]

Private individuals, corporations, municipalities and governments around the world also offered help. However, the New Zealand government turned down some of the most generous offers. Although Prime Minister George Forbes assured Morse that the government would do all it could, he turned down both a rehabilitation loan offered by the Bank of England at a peppercorn repayment, and an offer by the Mayor of San Francisco to raise a Mansion House Fund.

Total gifts of £396,000 ($28,512,000) were administered by a Relief Committee under the Public Trustee.[35] Some £241,000 ($17,352,000) of this went to repair 8500 private houses, though the criteria were strict. Householders had to come up with builders' quotes, and support per household was capped at £100 ($7200). The average pay-out was £28 ($2016), actual town averages varying from £15 ($1080) per house in Wairoa to £39 ($2808) per house in Napier.[36] Around 92 percent of all Napier householders appear to have received relief money.[37] However, this did not meet real costs. The Napier Municipal Reconstruction Department estimated actual repair costs for each house to average around £51 ($3672),[38] and at least 213 houses hit the £100 ($7200) ceiling.[39] David Dowrick, D.A. Rhoades, J. Babor and D.A. Beetham concluded in a 1995 analysis that total Napier housing repair costs were £145,648 ($10,486,656), or £64 ($4608) per house.[40]

On this basis, relief funds covered about 60 percent of the real cost,[41] a point which probably explains why some houses were never fully repaired — earthquake damage remained in one Napier South house even 70 years later.[42] Documents also paint a picture of downwards adjustment and walls of bureaucracy that angered and frustrated residents. Hastings resident Margaret Beamish was indignant about her treatment. 'The estimate sent was £33 7/- ($2401) which you have cut down to less than a fourth,' she wrote to the Reconstruction Officer in December 1931.

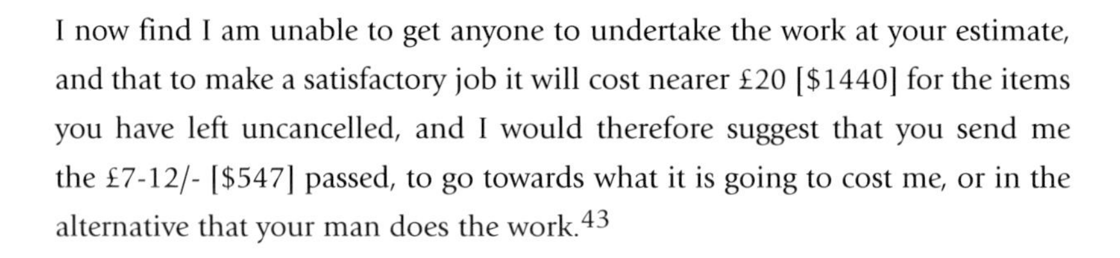

> I now find I am unable to get anyone to undertake the work at your estimate, and that to make a satisfactory job it will cost nearer £20 [$1440] for the items you have left uncancelled, and I would therefore suggest that you send me the £7-12/- [$547] passed, to go towards what it is going to cost me, or in the alternative that your man does the work.[43]

Others had trouble getting anything — irregular forms prompted irritating volumes of paperwork. 'With regard to your application for financial assistance,' the Reconstruction Officer wrote to Mrs A. Beattie of Meeanee in August 1931, 'you have omitted to sign the form at the foot ... Also fill in the Mortgagee's name and address, and the correct amount.'[44] This probably helped everyone get a reasonable share of the funds, but it did not endear the committee to the public.

Even when the bureaucratic hurdles were overcome, tradesmen were in huge demand and there were often long delays before houses could be repaired. Then they had to be inspected before they could be reoccupied. For many families this meant long separations. Certainly in the first weeks, women and children stayed out of the district — mostly in Wanganui or the Manawatu — while their husbands lived in temporary accommodation in Hawke's Bay and tried to get their homes repaired. Dorothy Campbell went to her family in Feilding. 'Our house was fortunately insured against earthquakes,' she wrote:

> ... but we don't know what we are getting out of it yet & can make no plans until we do know. In the meantime Ann, nurse & I will stop here & Neil over there coming here as often as he can while I will do the same. It will be hard to be separated but it seems to be the only way as Ann could not possibly live in the woolshed.[45]

Separation added to the strains and stresses faced by the survivors. Havelock North refugee E. Nicholson Fuller wrote to the Town Board requesting a special permit to return. Worried about her husband and son who had remained in the village, she asked to 'return to look after him'.[46] Even when families were reunited, many had to wait months before they could return to their houses. Mary Hunter and her husband did not finally move back into their Napier Hill flat until January 1932. 'Never were two people more glad to get home,' she wrote.[47]

William Edward Barnard (1886–1958), lawyer, politician and MP for Napier 1928–43, seen here in 1930. (S.P. ANDREW COLLECTION, PACOLL-3739, F-18364-1/1, ALEXANDER TURNBULL LIBRARY)

A further £95,000 ($6,840,000) of the donated relief money went to make good the accommodation, food, clothing and medical costs incurred in the wake of the disaster. Some £47,000 ($3,384,000) went as ‘permanent provision’ for the injured and their dependants, and £8000 ($576,000) went on ‘personal losses’. Two thousand pounds ($144,000) was given to ‘elderly persons unemployed as a result of the earthquake’. This left £3000 ($216,000) undistributed at the end of 1932.[48]

The new Napier

Reconstruction in the town centres looked likely to take years. Meanwhile, shopkeepers could not afford to remain closed and planning quickly got under way to provide temporary premises. A few structures were put up in Napier amid the piles of debris, and after some debate corrugated-iron shops were built in Clive Square. ‘Tintown’ flourished for several years before being torn down as the new town centre grew.

The destruction in central Napier was so complete that some hoped to wipe the slate clean, and there was discussion of a new town centre on the other side of the hill. In the end, though, Alfred Domett’s colonial design survived with a little widening and splaying on 87 corners.[49] A few thoroughfares were added.[50] R.S. Munro was of the opinion that even this might have been difficult to achieve under ordinary circumstances — the destruction of recent surveys produced disputes which might have dragged on if left to ordinary bureaucracy.[51] The extraordinary powers given to the Commissioners allowed them to force the point, and some £30,000 ($2,160,000) was eventually paid to property owners by way of compensation.[52]

What to build was another matter. Less than a fortnight after the quake the *Daily Telegraph* called for a new Napier along the lines of Santa Barbara in

Storkey’s booksellers and stationers recommenced business with this line of souvenirs; owner W.E. Storkey purchased photographs of the quake and had them published for curious locals and tourists alike.
(Storkey Collection)

With debris still piled high, corrugated iron provided shopkeepers with temporary premises in the weeks after the quake. Few could afford to remain closed for long in the depths of the worst depression the Western world had yet seen. Seventy-odd shops later opened in temporary corrugated-iron buildings on Clive Square.
(H.N. Whitehead Collection, PACoLL-3068, Alexander Turnbull Library 21336-1/4)

California. This had been completely rebuilt in Spanish Mission style after a devastating earthquake in 1925.[53] The newspaper's ideas were supported by Napier architects, and plans were floated under the guidance of a Reconstruction Committee. Free-flowing concepts included a single large Spanish Mission building covering the Tennyson–Emerson Street block. However, Napier during the depression did not have the resources of Santa Barbara at the height of the 'roaring twenties' and nothing on this scale was built, though proposals for grand structures bubbled along for years. When thoughts turned to developing the Marine Parade in 1933, J.A. Louis Hay proposed a huge street-spanning Entertainment Centre. This proved too expensive, and his magnificent Municipal Theatre — another creation in unashamedly modernist stylings — was also stillborn.

Reconstruction when it came was mainly in the form of accelerated town renewal on existing sites and mostly took place between late 1931 and early 1935, with a particular focus on getting the shops open. Neither the rate nor the volume of the work should be exaggerated.[54] Real progress had to wait for funding and new regulations. The Building Construction Act, passed in response to the earthquake, imposed uniform codes on new construction nationally from the beginning of 1932. It was a shift of emphasis rather than a complete innovation. Reinforcing had been a feature of pre-quake construction. The Act provided standards and laid the groundwork for further improvement, though the special fee attached to new permits to fund research into earthquake resistance raised howls of protest.

The first new downtown Napier structure was the steel-framed Market Reserve building. Designed by Natusch and Son, this building had been authorised by the council 'some months before the quake', and would have gone up during 1931 in any case.[55] Most privately owned structures went up as owners could afford to rebuild, their size and features constrained by the realities of cost. Architects from Hastings, Napier and Wellington combined to meet the workload. Members of 'Associated Architects' included J.A. Louis Hay,

The main trading banks clubbed together soon after the quake to build joint premises, known as 'community bank' buildings. The spire of the historic Catholic Church rises behind. This church – built in wood – survived the quake where brick-built churches did not. It was destroyed by fire in 1981.
(KITTY WOOD COLLECTION, PA COLL-1009-05, ALEXANDER TURNBULL LIBRARY)

J.T. Watson, H.G. Davies, A. Garnett and Eric Phillips. They coordinated their work as closely as possible — among other things, developing a consistent verandah height for the downtown shopping areas.

Almost all the new buildings adopted the austere styles in vogue during the early 1930s, making the town centres of both Napier and Hastings a small-scale snapshot of the architectural thought of the period. The slicker 'streamline modern' of the late 1930s was less in evidence, though W.J. Prouse's Masonic Hotel harked forward to this as early as 1932. Designs were eclectic. Wellington-based architects Crichton, McKay and Haughton adopted Maori motifs on the Bank of New Zealand. Mayan themes featured in the Haynes building further down Hastings Street. Architect E.A. Williams put sunburst and octagon motifs into the Central Hotel, which he designed for the Napier Brewery Company. J.A. Louis Hay remained loyal to the Chicago school and Frank Lloyd Wright, working their themes into the National Tobacco Company building at Ahuriri, among others.

Not every structure was rebuilt wholly from scratch. In Napier the Post Office, Public Trust, E & D and Dalgety buildings were reconditioned, and the shell of Parker's Chambers was partly recoverable. A few pre-quake structures were rebuilt to the original plans, including the Women's Rest in Clive Square, reconstructed in 1934.

FAR LEFT: The Market Reserve building was the first new downtown Napier building to go up after the quake. Authorised in December 1930, it would have been built regardless of the disaster and technically predated the post-quake reconstruction. Cantilevered verandahs were a particular feature of all post-quake buildings. New interest in Napier's heritage by the late twentieth century prompted careful restoration, seen here in this picture taken in early 2000.
(MATTHEW WRIGHT)

NEAR LEFT: The interior of the *Daily Telegraph* offices was characteristically modern: simple, vaulted and skylighted with fittings to the very latest styles. This original open fit-out was later hidden by partitions as demand for office space within the building grew.
(*DAILY TELEGRAPH* COLLECTION)

Reconstruction flourished in Napier particularly during the 1932–35 period. Most of the downtown shops were ready by 1933, and the ceremonial rebirth of Napier was celebrated during festivities that January. However, while reconstruction in Napier, Hastings, Havelock North and the other devastated centres doubled the region's usual slice of the national construction cake, this represented an increase only from 6 to 13 percent of the national total. It had dropped back to the usual 6 percent by 1935.
(EVENING POST COLLECTION, ALEXANDER TURNBULL LIBRARY C-24714-1/2)

All this work doubled the Hawke's Bay share of the national construction cake during the 1932–34 period, though this still accounted for only a small fraction of the New Zealand total. Specifically, Hawke's Bay's share of national building expenditure rose from 6 percent of the national figure in 1930 to a peak of 13 percent in 1934, before dropping back again to 6 percent.[56] Some 586 permits were issued for new buildings,[57] with a value in the two years to the end of March 1933 of £612,000 ($44,064,000).[58] These included £122,857 ($8,845,704) on new buildings at Ahuriri.[59]

Most of Napier's downtown shops were open by late 1932, and although much remained to be done, Napier was officially 'reborn' during January 1933 in a 'New Napier' celebration. The streets had been widened, some 25.5 miles (41 km) of new sewers had been laid, 2473 service drains relaid, and five new pumping stations built. New water systems included a reservoir holding 1,300,000 gallons (about 6 million litres), artesian bores able to deliver 4,000,000 gallons (over 18 million litres) a day, and repairs to 35 miles (56 km) of water mains and 2000 connections.[60]

Reconstruction continued for years, leavening the early 1930s look with later styles, as adopted by Atkins and Mitchell for their 1936 T & G building on the corner of the Marine Parade and Emerson Street. After much debate the new Municipal Theatre was built to a 1937 design by J.T. Watson. It featured zig-zag

One of the most impressive of the new Napier buildings was designed by E.A. Williams in 1932 for the *Daily Telegraph*. The primary office and printing works of Napier's long-standing daily paper included characteristic zig-zag motifs, repeated in ironwork around the doors.
(*Daily Telegraph* Collection)

The Criterion Hotel on the corner of Hastings and Emerson Streets went up after the adjacent Market Reserve building. Heavy framing and reinforcing in this Spanish Mission structure is evident in this picture; afterwards, the building proved extremely difficult to alter.
(Crompton-Smith Collection, PAColl-4073, Alexander Turnbull Library F-28216-1/2)

motifs and interior side-lights connected by streamline shapes. Watson was also responsible for some of the Marine Parade developments completed towards the end of the decade, including the 'New Napier Arch' of 1940, adjacent to the memorial and soundshell. The soundshell itself, built with funds raised by the Thirty Thousand Club, was copied from a similar structure in Hollywood. The complex included a semi-circular earthquake memorial, featuring HMS *Veronica*'s bell.

A few structures were not replaced for decades. Although a wooden church was built on Napier's Anglican cathedral site in 1932, a permanent replacement was only begun in 1955. It was dedicated in March 1965.[61] A new nurses' home waited many years, until a multi-storey structure finally went up in the 1950s. The biggest slip on the face of Bluff Hill was not cleared until the 1960s.

Napier's Art Deco Heritage

The architectural stylings of the twentieth century were already established in Hawke's Bay when the quake struck. Some of the newer buildings reflected the 'decorative arts' movement; others harked across the Pacific to the Californian Spanish Mission style. However, early plans for a Spanish Mission theme fell victim to depression-era economics. What followed was an accelerated urban renewal, mostly on existing titles, producing a unique concentration of small art-deco and Spanish Mission buildings. The heritage value of these structures — inevitably — slipped the attention of the generation that built them, but interest was revived in the mid-1980s on the back of a general resurgence of early twentieth century stylings. By this time some of the best post-quake buildings had already fallen victim to urban renewal. The rest were preserved, refurbished, and became the focus of local interest in all things art deco, as well as a drawcard for tourists.

Wellington architects Crichton, McKay and Haughton designed the Bank of New Zealand building on the corner of Hastings and Emerson streets in 1932. It featured Maori motifs and was one of three modernist-styled banks erected on the Hastings and Emerson streets intersection. By the early twenty-first century it was the sole survivor of the trio, and premises of the Auckland Savings Bank.
(Matthew Wright)

Spanish Mission and two eras of art deco; the Provincial Hotel stands to the left, designed by Finch and Westerholm in 1932. A 1980s emulation of the style intrudes centre-frame.
(Matthew Wright)

Napier's art deco revival of the 1990s transformed the town centre; old buildings were restored and reframed with modernist-style plants and street lights. This is the former Hotel Central, a 1931 design by E.A. Williams.
(Matthew Wright)

Detail of the former Hotel Central.
(Matthew Wright)

Some of Napier's modernist-era shop frontages survived into the 1990s, restored with the art-deco revival of the day.
(Matthew Wright)

Details of Napier's art deco buildings.
(Matthew Wright)

New complements old, the Pan Pac foyer of the Napier Municipal Theatre was a late 1990s addition to J.T. Watson's 1937 design.
(MATTHEW WRIGHT)

J.T. Watson's New Napier Arch of 1940 symbolised the rebirth of the town.
(MATTHEW WRIGHT)

Iconic Napier art deco; the Masonic Hotel, designed by W.J. Prouse in 1932.
(MATTHEW WRIGHT)

The Napier war memorial, opposite Clive Square, with the Women's Rest in the background. 'Tintown' stood here in the months after the quake. The Women's Rest was a pre-quake design of J.A. Louis Hay, rebuilt to the same design in 1934.
(MATTHEW WRIGHT)

Rubble from the quake was pushed to the town beach, forming the foundation of this wide public area, which was initially used as a skating rink. The Sound Shell dominates the centre of the frame with the Sun Bay quake memorial and home of the *Veronica* bell to the left.
(MATTHEW WRIGHT)

The New Napier celebrations of January 1933 marked the official rebirth of the town centre, though much remained to be done. In this view of Emerson Street, looking towards the sea, most of the shops have been finished; but the Hastings Street–Marine Parade block remains empty and work has yet to begin on the Marine Parade beyond. The single-storey shop second from the left is Blythes.
(*Evening Post* Collection, Alexander Turnbull Library C-21512-1/2)

By the early 1960s business was booming and the post-quake buildings with their cantilevered verandahs were already showing their first signs of age. This view down Hastings Street begs comparison with the centre and bottom images of the fire raging up the same stretch of street on page 51. Thorp's shoe shop is visible on the left.
(*Daily Telegraph* Collection)

The multi-storey Ward Block addition to the Napier Public Hospital went up on the site of the original 1930 nurses' home in the late 1960s. By the time this picture was taken in late 2000, the Napier hospital had been closed in favour of expanded facilities in Hastings.
(Matthew Wright)

REBUILDING HASTINGS

Although Hastings had suffered less complete devastation than Napier, much still had to be done to restore the town. A 1998 study revealed that only six of 142 unreinforced buildings survived the quake undamaged. Five out of 36 reinforced masonry buildings collapsed and a further fifteen were cracked.[62] Afterwards, 89 buildings were wholly condemned, and parts of 79 also had to be demolished.[63] H.F. Baird inspected many buildings in the days after the quake and worked out what had happened:

> Many damaged brick walled buildings had been bedded on but a shallow foundation of poor concrete, they had carried, by loose inter-connection, poorly tied roof trusses with heavy roofs of wide spread. In some two storey buildings of this type top floors just as poorly connected to the walls had been carrying enormous loads of merchandise. As soon as these buildings were subjected to horizontal accelerations at ground level, the transmission of vibrations to the superstructure … caused disruption and ultimate collapse. The shallow bed of concrete was a weak point too … The roof ceiling beams, cantilever verandahs, and so forth, had tended to burst walls of poor coherence. Beams in stairs and various joists could even have acted as battering rams in discontinuities of design.[64]

The first priority was premises for hard-pressed retailers. Some 194 permits for temporary buildings were issued, with a value of £30,000 ($2,160,000). 'When you come home,' Bill Ashcroft wrote to his sister Catherine, who was training in England, 'you won't know Hastings as they are only allowing tin sheds to be built for 12 months until they decide what building regulations to make for earthquake proof buildings.'[65] In the event some owners felt their 'tin sheds' were good enough — permits were reviewed in August 1933 in an effort to get rid of some temporary buildings. Fourteen were reclassified as permanent, but as late as 1939 there were still nine temporary buildings in the borough. A further 422 permits were issued for £341,000 ($24,552,000) worth of new buildings.[66]

For Hastings businessmen the months after the earthquake were dominated by hot debate over widening Heretaunga Street. Internal Affairs Department town planner J.W. Mawson set the cat among the pigeons in April when he told a public meeting that new buildings should be set back five feet (1.5 m) between Willowpark and Tomoana Roads. Ex-mayor George Ebbett, who owned property in the street, petitioned the council to stop the widening. Mayor G.F. Roach set his new temporary premises back five feet in anticipation

By late 1931 all of the loose rubble had been moved in Hastings, but broken buildings still featured in the centre of town. Reconstruction was delayed by a wrangle over whether to widen the main street. Temporary buildings, as seen here on the left-hand side of the street, were a common sight for many years. Westerman's, on the right-hand side of the picture, was demolished soon after this photograph was taken and replaced with a Spanish-inspired structure designed to the latest styles by Edmund Anscombe.
(*Evening Post* Collection, Alexander Turnbull Library F-103402-1/2)

As in Napier, reconstruction in Hastings was a relatively slow affair which had to wait on funding. Individual property owners quietly refurbished or rebuilt their premises as time went on.
(H.N. Whitehead Collection, PAColl-3068, Alexander Turnbull Library F-19296-1/1)

It was silly nose day for Joseph Nathan & Co. when their new premises opened in 1932. The building harks forward to later 'streamline modern' and incorporates a number of distinctive features, including zig-zag pediments and stepped decorations.
(H.N. Whitehead Collection, PAColl-3068, Alexander Turnbull Library G-4791-1/1)

and launched a counter-petition. It turned out that 64.9 percent of ratepayers wanted the street widened. Although property owners led by Ebbett voted 21–6 against, the council decided to adopt the scheme. Mawson arranged valuations, but Ebbett was far from sunk and by the end of July had persuaded the council to drop the idea.[67]

Rebuilding began in earnest during 1932 and continued for several years as individual property owners obtained funding. At least 20 local builders contributed. Much of the work in Hastings was done by Fletcher Construction at special rates, via their local agent H.W. Abbott. Reconstruction was again a snapshot of early 1930s styling, mixed with older Spanish Mission themes, and as in Napier some earlier structures were built anew, including Garnett's damaged 'Villa d'Este' which was demolished and reconstructed to the original design.

The first large building, Westerman & Co.'s £11,300 ($813,600) two-storey shop on Heretaunga Street, was designed by Anscombe in updated Spanish Mission style. By agreement, the next-door Harvey's Building of 1933 also had heavy Spanish Mission influence. Joseph Nathan & Co.'s new building of the same year was pure modernism. So too were the Garnett-designed Holden's building of 1934, Nutter's building on the corner of Karamu and Heretaunga Streets, and the Hurst building. The State Theatre was another period masterpiece, featuring characteristic 'winged' horizontal decoration. G.F. Roach's new shop of 1934, on the corner of Heretaunga and King Streets, featured second-floor streamline motifs topped with a low turret. Eric Phillips

The New Zealand Loan and Mercantile building was also a masterpiece of 1930s style, one of several distributed around Hastings between older Spanish Mission structures.
(GORDON BURT COLLECTION, PACOLL-4118, ALEXANDER TURNBULL LIBRARY F-37040-1/2)

and Harold Davies designed at least ten shops and organised the restoration of buildings such as the Pacific Hotel, the Assembly Hall and the Municipal Theatre. Their other work included the Commercial Bank of Australia's premises of 1932, and the Dental and Medical Chambers of 1935. S.G. Chaplin won a 1934 competition to design a new town clock, which went up adjacent to the railway line the following year and included bells rescued from the collapsed post office tower.[68]

Hospital services attracted considerable debate. Government architect J.T. Mair and F.W. Furkert, Engineer in Chief, inspected Napier Hospital as early as 10 February, reporting that the buildings had 'stood a battering far in excess of anything experts in design of earthquake resisting buildings consider maximum', yet Coleman and Midgeley Wards, along with the operating theatres, were never 'near point of collapse'.[69] However, the pro-Hastings lobby pointed out that the earthquake had underlined the need for a full hospital in Hastings. Funds were available in the form of a bequest of £37,000 ($2,664,000) from the estate of Henrietta Kelly, who had been killed in the earthquake. These arguments undermined support for Napier Hospital to the point where reconstruction funding was slashed from £110,000 ($7,920,000) to £66,000 ($4,752,000) in July 1931. Reconditioning of the surviving wards nevertheless got under way, and the hospital reopened in January 1932. Meanwhile, £5000 ($360,000) was made available in September 1933 from the Kelly bequest to bring Hastings Memorial Hospital up to 50 beds.

Hastings' architectural heritage underwent a renaissance during the 1990s. In this picture, taken early in 2000, Westerman's post-quake Spanish Mission-influenced building has been immaculately restored. (MATTHEW WRIGHT)

The expression of various 'modernist' styles in Hastings was not always apparent against the better known backdrop of Spanish Mission, but included some excellent detailing.
(MATTHEW WRIGHT)

REGIONAL RECONSTRUCTION

There was a good deal of damage around Hawke's Bay. H.F. Baird 'managed to get to Puketitiri with Mr Douglas Lane who was going up in his car' a few days after the quake. He discovered that the damage seemed to decrease between Taradale and Hakowai, and 'from Hakowai to Rissington the damage occurred in patches, road fillings suffered most near Rissington, and except for minor increases on cuttings and spurs, the damage tapered off towards Puketitiri'. On the way back he noticed fissures near Waiohiki which:

> ... ran along gullies at the foot of the hills, it seemed practically continuously for miles towards NE ... On the hill overlooking the golf course a substantially built wooden house of two storeys had its chimneys much damaged, while tanks had been thrown to the south. The verandah posts in front had been canted to the west. The timber in the house was very sound and extensive inside damage appeared to have been done almost entirely by the falling of the central chimneys which were very tall ... Near Hastings low paddocks were seen with water lying in them. There had been no rain since the earthquake. The driver said that before the earthquake these paddocks had been dry, on the previous day the pilot had made similar comment. The layers below were relatively impervious to water though it had been forced through by the earthquake.[70]

About 50 bridges needed work, and while some were promptly repaired by Hawke's Bay's various county councils, money was short and other projects languished, among them a new bridge between Napier and Westshore. It was 1932 before the embankment bridge could be repaired, spurring calls from Westshore residents for direct access over the harbour to Napier. They had to fight this to ministerial level before the Hawke's Bay County Council relented and built a footbridge. This opened in 1935; residents had to wait until 1961 for a road bridge.

The seven-year-old Dartmoor bridge, constructed in concrete at a cost of £3000 ($216,000), was left impassable — one of the spans tilted down into the river. It took a settler petition in 1932 to get a low-level temporary bridge. A permanent bridge was begun the following year, but delayed by dispute and then destroyed by the floods of 1938. A permanent replacement was finally completed in 1954.

The earthquake laid waste to river protection schemes. Long-standing argument over what to do about Heretaunga's wandering rivers had been resolved in 1930 with a compromise decision to split the Tutaekuri between an overflow channel to Waitangi and the main bed into the Ahuriri harbour. The quake wrecked that approach, knocking down essential stopbanks and raising the Napier end of the Tutaekuri by over three feet (about a metre). County Council officials were particularly worried about the vulnerability to flood of the emergency hospital at the Greenmeadows racecourse. Prime Minister

Westshore residents had particular problems reaching Napier, just over the harbour. Although part of Hawke's Bay County when the quake struck, Westshore was incorporated into the Napier borough in 1942, but a direct road bridge had to wait until the early 1960s. (STORKEY COLLECTION)

Forbes offered a £10,000 ($720,000) grant for immediate repairs as early as 11 February.[71] Three hundred men and 200 horses laboured to restore the stopbanks before the autumn rains, and managed to get the work done in a month.

Finding a longer term answer was complicated by the way the quake had turned the Ahuriri lagoon into a swampy morass. W.E. Barnard suggested that the Harbour Board should claim this 'gift from the sea'. Crown solicitor C.H. Taylor agreed. However, nobody could agree on how to turn it into farmland. Early plans suggested diverting the Tutaekuri into it as a source of silt, but in October 1931 engineer Guy Rochfort suggested that the Tutaekuri should instead be diverted to the sea down the existing overflow channel. The lagoon could be reclaimed with silt-heavy water flowing from the surrounding hills.

The issue was formally debated a few months later. The Harbour Board liked Rochfort's scheme but needed more funding. F.C. Hay proposed an amended scheme with an estimated cost of £270,000 ($19,440,000). The projects were authorised by the Napier Harbour Board Empowering Act and the Hawke's Bay Rivers Amendment Act of 1933. Work began on the Tutaekuri in March 1934, using local labour under Public Works Department supervision. By mid-1936 work had also got under way on the Ngaruroro. The final cost topped £360,000 ($25,920,000), but the scheme ended decades of bitter political debate over the future of Hawke's Bay's river systems.

DEFLECTING THE DEPRESSION

The earthquake came as New Zealand was descending into the worst depression the industrial world had known, during a summer drought that crippled the horticultural industry on which Hawke's Bay relied.[72] It was a triple blow — yet, although partial loss estimates were assembled as early as 1933,[73] no proper analysis was made of the economic impact until 1997 when economist Simon Chapple analysed the figures in some detail and concluded that 'a Keynesian model of activity' fitted post-quake developments better than the neo-classical theory in vogue by the 1990s.[74]

The question that must follow is whether reconstruction spending saved Hawke's Bay from the worst effects of the depression. Published statistics paint a dismal picture. Post-quake reconstruction generated extra income equal to about 2 percent of pre-quake figures.[75] About 300 jobs were created, principally in building construction.[76] However, this had no effect on the regional share of unemployment, which climbed from 1.5 to just over 2 percent of the national total between mid-1930 and mid-1933.[77]

The cost of repairing provincial bridging prompted hot political debate after the quake. By 1931 this bridge over the Tutaekuri at Redclyffe was 50 years old and long overdue for replacement. A new stressed concrete structure had been approved in 1925 but been delayed by strikes, funding shortfalls and problems with delivery of steel joists from England. It had yet to be started when the earthquake felled the venerable old wooden bridge. The new bridge finally opened in 1934. (KITTY WOOD COLLECTION, PA-COLL-1009-03, ALEXANDER TURNBULL LIBRARY)

What happened is clear from the Hastings experience. There were 817 unemployed in the borough by April 1931. About half the 278 applications for emergency relief between March and July 1931 were attributed to the disaster. Despite reconstruction work the number of unemployed in the borough rose to 958 in July 1932. This was largely caused by migrants arriving in the hope of finding jobs. As early as mid-1931 there were 24 locals and 101 outsiders working on reconstruction in Hastings and Havelock North. This caused a good deal of resentment, and at a public meeting in June, registered local unemployed decided to boycott any firm that hired out-of-town labour.[78]

Demand for reconstruction labour had little effect on wages. There was a short boom in several Hawke's Bay sectors after the quake but, on average, wages generally fell by about half between 1926 and 1935. All this reflected wider depression figures — in other words, post-quake spending had no real impact.

Actual capital losses are another matter. Historians have variously estimated these at anything between £3,750,000 and £16,500,000 in 1931 values[79] — equivalent to between $270 million and $1188 million in late-twentieth- century money. These variances partly stem from the fact that no complete contemporary estimates were put together. Some losses were never quantified, and those that were often relied more on guesswork than hard data, or confused theoretical with actual replacement cost. The destruction of household breakables, for instance, was estimated at £156,000 ($11,232,000) by District Engineer A. Dennie; about £10 ($720) per house.[80] However, not all household losses were reported, nor was all lost property fully replaced.

The best analysis remains Chapple's, whose 1997 study identified losses of £395,000 ($28,440,000) for housing, £15,000 ($1,080,000) for cars, and £1,000,000 ($72,000,000) for business buildings from Napier to Woodville, including farms. Of the £1,900,000 ($136,800,000) worth of stocks on hand when the earthquake struck, Chapple argued, around £400,000 ($28,800,000) worth was destroyed, stolen or lost. Real local body losses — including the harbour repair figure — amounted to £527,000 ($37,944,000). Chapple added a further £552,747 ($39,797,784) for central government, coming up with a final figure of £3,402,500 ($244,980,000). Given the uncertainties, he suggested that capital losses could have been as high as £4 million ($288 million), concluding that contemporary estimates of around £3,750,000 ($270,000,000) were 'likely to be much closer to the mark' than later figures.[81]

Chapple's top figure of £4 million, which translates to about $288 million in late-twentieth-century money, is broadly comparable to the cost of the 1987 Edgecumbe shock which caused damage worth $315 million.[82] In other words — even given the uncertainties of the data — the Hawke's Bay earthquake of 1931 was demonstrably one of the costliest disasters ever to strike New Zealand.

CHAPTER 6

An unsteady legacy

The sheer size of the 1931 earthquake and the fact that it made a 'direct hit' on the main centres of Hawke's Bay set it apart from anything that had gone before. Massive destruction in such a well-populated area was unprecedented in New Zealand. However, exactly what happened — even the strength of the shake — was not unravelled until the late 1990s.

By 1931 all the major geological features around Hawke's Bay had been identified, along with the directions of uplift and movement, but the mechanisms remained unknown.[1] Geologists groped for explanations. A few days after the quake, noted seismologist Sir Edgworth David explained that it had been a result of ocean pressure. 'The ocean is getting deeper and deeper and exerts such a tremendous pressure on its abutments ... that on reaching the straining point they finally give way.'[2] By this time Alfred Wegener had proposed his theory of plate tectonics, but the inertia of 'progressivist' nineteenth-century thinking, coupled with the fact that Wegener hadn't identified the mechanisms behind plate motion, meant his ideas were not accepted by the scientific community until the 1960s.[3] As soon as the mechanisms were identified, though, plate tectonics became the only possible explanation.

Most New Zealand earthquakes result from interaction between the Indo-Australian and Pacific crustal plates, which float towards each other on a deep layer of plastic rock known as the asthenosphere.[4] East of Hawke's Bay, the Pacific plate dives below the edge of the Indo-Australian, into the asthenosphere, where it is eventually consumed. This process is known as subduction. The point of intersection is the Hikurangi trench. The result in Hawke's Bay is a series of fault lines.[5] Surveying by triangulation as early as the 1870s identified the general direction of movement, though the mechanisms were not understood. A formal geodetic survey of New Zealand between 1909 and 1949,[6] coupled with further work in the 1970s, provided more data. The shift averages 50 mm per year at the Hikurangi trench and 7 mm a year along the central ranges. Between the two, the hills in eastern Hawke's Bay — such as Te Mata Peak — are pushed up at an average rate of 2 mm a year.[7] Shells on the peak indicate that it was seabed as recently as five million years ago.[8]

WHAT HAPPENED IN 1931

The cause of the 1931 quake was discovered during the 1970s, when geologists found that the axis of tectonic movement in Hawke's Bay had shifted. Pre-1930 data showed that this ran in a northwest/southeast direction. By the 1970s, the movement was running northeast/southwest. This indicated what had happened. The 1931 quake was caused by irregularities in the surfaces of the colliding plates locking them together, skewing the forces into a new direction, and allowing energy to build up. This probably occurred before Europeans reached Hawke's Bay. The obstruction was overcome in February 1931 — releasing the pent-up energies in a cataclysmic jolt that created an abrupt upwards thrust of about 3 metres in a 100-kilometre fault,[9] not far northwest of Napier. Technically it was a strike-slip movement, with some reverse-slip components.[10] The earthquake relaxed seismic tensions and returned the direction of plate movement to normal.

It took years for seismologists to work out the details, mainly because of poor data. The New Zealand seismograph network was incomplete in 1931, and the shock overloaded most of the operating units. The only complete records were taken by an Imamura seismograph at Takaka and a Milne-Jaggar device in Wellington. The Takaka seismograph had only just been installed and was not properly calibrated, and neither device had photographic recording. Seismologists at the time noted that this was 'far from satisfactory'.

Initial calculations in 1933 put the epicentre at 39° 20'S, 176° 40'E — between Rissington and Patoka, northwest of Napier — and at an approximate depth of 'probably between ten and fifteen miles [16 and 24 km]'. Guthrie-Smith published a map of the Patoka location in *Tutira*.[11] But these were only estimates. In 1938 the figures were revised to 39.3° S, 177° E,[12] the rugged beach of Aropaoanui, 19 miles (around 30 km) from the originally estimated epicentre and much closer to the main centres of population.

A 1987 study found that the 'primary rupture surface' — the rock that moved — extended northeast and southwest of the epicentre, running from the seabed northeast of Napier southwest in a line past Hastings to Pukehou.[13] In this sense the earthquake scored a direct hit on Napier and near misses on Hastings, Havelock North, Wairoa, Waipawa, Waipukurau and the surrounding country districts. Near Napier there were ground accelerations of up to one gravity,[14] combined with lateral acceleration of up to a quarter of a gravity. Elsewhere, accelerations were less. These calculations were not far from 1931 estimates. H.F. Baird, examining the countryside around Hawke's Bay a few days after the quake, estimated that accelerations of up to one-third of a gravity could have caused the sort of damage he was seeing.[15]

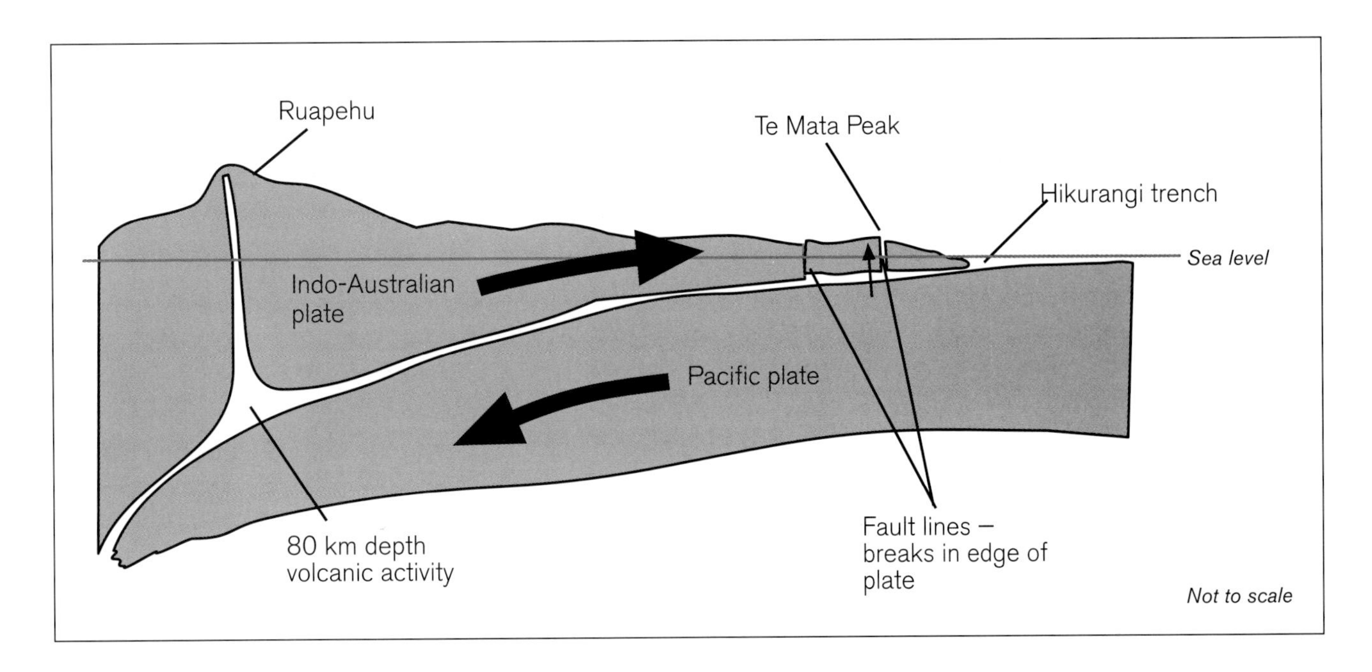

Indicative plate boundary showing Hawke's Bay.

The strength of the Hawke's Bay earthquake was also revised as time went on. Professor Charles F. Richter originally calculated the strength of the quake to have been 7.9 on his intensity scale, a figure he did not change when he revised his calculations in 1958.[16] By the late 1970s, however, Richter's concepts had been superseded by several systems that more accurately reflected seismic realities.[17] Perhaps the most meaningful was the measure of moment magnitude — M_w, the amount of slip along a fault. K. Abe concluded in papers published in 1981 and 1983 that the 3 February 1931 earthquake had a magnitude (M_w) of 7.8 — less than previously estimated. A further study by David Dowrick in 1990 confirmed these findings.[18] These revisions put the earthquake on a par with recalculated figures for the 1929 Murchison shock, and it became known as the 'Magnitude 7.8 1931 Hawke's Bay, New Zealand earthquake'.[19]

The 'felt intensity' — how an earthquake is felt in various places — is perhaps more useful than an abstract measure. A scale first defined by the Italian seismologist Mercalli was revised for New Zealand conditions by George Eiby in 1966, and modified again in the early 1990s on the basis of the Edgecumbe earthquake.[20] The Modified Mercalli (MM) strength of the Hawke's Bay earthquake was quantified by David Dowrick in 1998. His analysis revealed that in Napier, Hastings, Taradale and Petane the intensity peaked at X, while in Waipawa, Waipukurau and Wairoa it was as high as VIII. The peak intensity was as strong as the most violent earthquakes ever recorded in New Zealand.[21]

LEGACY OF 1931

The human toll was so horrific that new building codes were drawn up in 1931, specifically from the Hawke's Bay experience. These were applied nationally, and later revised to reflect new understanding of earthquake effects and new engineering techniques. Compulsory earthquake insurance began in 1944.

Fire was another issue of deep concern. In June 1931, J.S. Barton investigated the fires that had annihilated central Napier. Gas and power had been cut off by the quake, and 'an enquiry spread over three months ... failed to disclose in any residential or business premises in Napier any damage, even a scorched wall, as a result of gas flame'.[22] Barton suspected that broken jars had created a cocktail of 'fluids and vapours readily ignitable' in the pharmacies, possibly in the presence of 'naked gas flames', and recommended further research and experiment.[23] He also discovered that the salt-water sump in Clive Square was one of several sunk around the town for fire-fighting back in colonial days when the town water supply came in barrels. It turned out that many of the sumps had become blocked through disuse, and the brigade did not know where they all were. With thoughts of preventing post-quake fire calamities around New Zealand in the future, Barton suggested sumps should be dug in all towns 'near enough to the sea to make such a measure possible'.[24]

It did not take long for the wisdom of all these decisions to become evident. A 5.9 magnitude shock hammered Hawke's Bay in May 1932. Peak intensity in Napier and Taradale reached VI or VII on the MM scale. Fortunately, reconstruction had hardly begun and damage was minimal. A larger shock hit Wairoa and Gisborne six months later, peaking at between IX and X on the MM scale. Great damage was done to chimneys, buildings collapsed in Gisborne and Wairoa, and bridges were twisted. Seismologists later concluded that this had probably been in the same rupture zone as the 1931 shock.[25] The region was shaken again in March 1934 by a magnitude 6.4 shock that had a peak intensity north of Wairoa of VI on the MM scale.[26] Two more shocks whacked into southern Hawke's Bay in December 1938.

Perhaps the closest Napier came to a repeat of 1931 was on 6 October 1980. The shallow quake that struck at 3.30 a.m. had a peak magnitude of 5.9, and was located so close to Napier that the intensity was V or VI on the MM scale. Some people woke before it struck, and in the stillness of the night the roar of the approaching shock wave was clearly audible. Groaning walls bent like reeds in a gale as it rolled past,[27] jolting everyone awake from Wairoa to Waipawa. Goods were hurled from shop shelves in Napier and Hastings. Some windows broke. Trains could not run until lines were checked, lifts had to be inspected, and the Earthquake and War Damages Commission fielded 2000 claims.[28]

A magnitude 5.8 quake hit Hastings in 1982. Dannevirke and central Hawke's Bay were racked by two major shakes in 1990.[29] Although these caused a good deal of damage, buildings remained upright and there were no deaths. The lessons of 1931 had been well and truly learned. National pay-off came in 1987 when a major shock hammered Edgecumbe. Material costs were in the same league as the 1931 Hawke's Bay shake — after correcting for inflation — but nobody died directly from injuries.[30] In short, the 1931 quake provided an object lesson which New Zealand quickly learned. Formal earthquake planning, building regulations, and mechanisms to fund recovery after natural disaster can largely be traced back to the lessons learned in Hawke's Bay during 1931.

Yet the response in 1931 was by no means slow or ineffective. Indeed, the efficiency with which local and national authorities improvised relief measures was truly outstanding. Despite massive damage to the infrastructure — including the total destruction of the only base hospital in the district — food, water, medical care and evacuation were available for victims within 24 hours. Power, railway and telegraph were essentially restored within 36 hours. All this was done without significant prior planning — a remarkable achievement that demands explanation.

To some extent luck played a part. It was fortuitous that the *Veronica* was in Napier and the two cruisers were available to leave Auckland within hours. However, the Navy would have mobilised immediately to help in any case, and at this level the response was a function of the resources available. The problems of under-resourcing were apparent from NZPAF experience — the younger organisation had to borrow aircraft to offer even minimal aid. The early restoration of essential services, particularly power, was a function of the environment of 1931. The national grid was designed and built as a unified and integrated entity, which made it possible to quickly jury-rig connections and draw on national resources. This would not have been so easy if the grid had been a fragmented jumble of small competing suppliers. Similarly, railways and essential services such as telegraph and telephone were publicly owned resources whose first responsibilities lay with the public. These government organisations were immediately able to draw on a national support base without considering the impact of cost on their end-of-year profits. These characteristics facilitated the recovery of service in the devastated areas and helped ease the plight of those caught in the disaster.

Credit for the rapid response to disaster must also go to the people of the day, and the society that flourished after the First World War. New Zealand adults in 1931 had either served in or lived through the war. Their experience laid important groundwork for post-quake action, both in terms of command structures — particularly in Hastings where one wartime military unit was actually resurrected — and in the broader community bonds forged in the field of battle at Gallipoli and Flanders. This resurfaced in the days after the earthquake. 'Too much cannot be said of the splendid spirit of the Napier people,' one reporter wrote. 'There is something in big catastrophes that brings out the best in everyone, as did the Great War, and the way in which people are forgetting themselves to others strikes the stranger forcibly. Property is nothing. Fatigue and personal feelings do not count.'[31] The comparison with wartime experience was frequently and consciously made. 'Scenes reminiscent of the war time in France,' read one caption in the *Daily Telegraph*'s 1931 book *Before and After*.[32] One contemporary account wrote glowingly of the way 'neighbour helped neighbour', and 'for the time being, at least, there is a greater human understanding of each other than ever there was before. May it continue!'[33]

Under these circumstances it was natural for ordinary people simply to do what they could to help, regardless of cost or their own losses. Many of those in positions of authority worked on for hours and days despite personal loss, among them Constable Tripney who knew his wife and infant son had been killed, but 'continued to work … with rescue and search parties'.[34] Doctors and nurses such as M.D. MacNab volunteered their services without delay or question. It was equally natural for miscreants such as Barggren to do what they could to help during the disaster and then turn themselves in again. These were not unusual acts of altruism: they were normal responses to disaster from a society that genuinely cared.

Notes

Abbreviations

EPR Earthquake Personal Reminiscences
HDC Hastings District Council
HNL Havelock North Public Library
HPL Hastings Public Library
NAMU Hawke's Bay Museum Library
WTu Alexander Turnbull Library

Introduction

1 WTu, MS Papers 2418, Folder 4. W.J. Ashcroft writing of the quake in 'The Apiarist' in *The New Zealand Smallholder*, 16 March 1931.
2 *New Zealand Official Yearbook 1931*, p. 96, estimated Hawke's Bay's 1930 population to be 69,500.
3 'The Full Story of the Great Earthquake Disaster', *The Weekender*, Third Overseas Edition.
4 J.G. Wilson, *History of Hawke's Bay*, p. 451, cited 246 casualties, possibly a typographical error for 256. A.H. McLintock (ed.), *An Encyclopedia of New Zealand*, Government Print, Wellington, 1966, Vol. 1, p. 475, cited 256 casualties, 161 in Napier, 93 in Hastings and 2 in Wairoa. This is the official tally. Geoff Conly, *The Shock of '31*, A.H. & A.W. Reed, pp. 232–235, listed 258 names comprising 140 in Napier and 22 unidentified, 87 in Hastings and 6 unidentified, and 3 in Wairoa.
5 *Hawke's Bay — Before and After, Daily Telegraph*, Napier, 1931, reprint 1981, p. 66.
6 The most lethal single events in New Zealand between 1840 and 2000, other than the Hawke's Bay earthquake, were: Mt Erebus DC-10 crash 1979 —257; Tangiwai train crash 1953 — 151; *Penguin* sinking 1909 — 75; *Wahine* sinking 1968 — 51. The two world wars and the 1918–19 influenza epidemic were more lethal, but took place over a period of time.
7 *New Zealand Herald*, 5 February 1931.
8 Simon Chapple, 'The Economic Effects of the 1931 Hawke's Bay Earthquake', New Zealand Institute of Economic Research (Inc.), Working Paper 97/7, Wellington, August 1997.
9 David Dowrick, 'Damage and intensities in the magnitude 7.8 1931 Hawke's Bay, New Zealand earthquake', *Bulletin of the New Zealand National Society for Earthquake Engineering*, Vol. 31, No. 3, September 1998.
10 Eric Hobsbawm, *On History*, Abacus, London, 1998, p. 284. Hobsbawm's italics.

Chapter 1: 'A bolt from the blue'

1 WTu, MS Papers 2418, Folder 4. W.J. Ashcroft writing in 'The Apiarist' in *The New Zealand Smallholder*, 16 March 1931.
2 WTu, MS Papers 3359. *Taranaki* (ship).
3 *Hawke's Bay — Before and After*, p. 78.
4 WTu, MS Papers 5814. Dorothy Campbell, letter to 'Aunt Frances', 5 March 1931.
5 David Thorns and Charles Sedgwick, *Understanding Aotearoa/New Zealand: Historical Statistics*, Dunmore Press, Palmerston North, 1997, p. 74. Peak depression unemployment rate in 1936 was 8.7 percent or 55,505 individuals. In 1992 the rate was 11.2 percent or 172,600 individuals.
6 R.J. McDougall, *New Zealand Naval Vessels*, Government Print, Wellington, 1989, pp. 18–19.
7 *Hawke's Bay — Before and After*, p. 85.
8 Matthew Wright, *Hawke's Bay — The History of a Province*, Dunmore Press, Palmerston North, 1994, p. 155.
9 NAMU, Earthquake Reminiscences Box 1 (EPR), Bernard Chambers Memoir (diary extracts).
10 Dowrick (1998).
11 Matthew Wright, *Havelock North — The History of a Village*, HDC, Hastings, 1996, p. 174.
12 WTu, MS Papers 5814. Dorothy Campbell, letter to 'Aunt Frances', 5 March 1931.
13 See, e.g., Jefley J. Aitken, *Rocked and Ruptured: Geological Faults in New Zealand*, Reed Publishing (NZ) Ltd, Auckland, 1999, p. 32.
14 D.J. Dowrick, D.A. Rhoades, J. Babor, and R.D. Beetham, 'Damage ratios and microzoning effects in Napier in the magnitude 7.8 Hawke's Bay, New Zealand earthquake of 1931', *Bulletin of the New Zealand National Society for Earthquake Engineering*, Vol. 28, No. 2, June 1995, p. 134.
15 EPR, Mitchell des Landes memoir.
16 WTu, MS Papers 2418. W.J.C. & W.H. Ashcroft Collection, W.H. Ashcroft to his daughter, 15 February 1931.
17 Ibid.
18 EPR, Jessie Atkinson memoir.
19 *Hawke's Bay — Before and After*, p. 75.
20 Ibid., p. 80
21 EPR, W. Olphert memoir.
22 *New Zealand Herald*, 6 February 1931.
23 *Hawke's Bay — Before and After*, p. 99.
24 Ibid., p. 156.

25 *New Zealand Herald*, 5 February 1931.
26 WTu, MS Papers 2107. Jean Anderson, letter to Elsie Young, 3 February 1931.
27 *Hawke's Bay — Before and After*, p. 63.
28 WTu, MS Papers 2107. Jean Anderson, letter to Elsie Young, 3 February 1931.
29 WTu, MS Papers 2418, Folder 1. W.J.C. & W.H. Ashcroft Collection, Bill Ashcroft to his sister, 15 February 1931.
30 *New Zealand Herald*, 5 February 1931; also *Hawke's Bay — Before and After*, p. 66.
31 J. Wright, pers. comm.
32 *Hawke's Bay — Before and After*, p. 68.
33 *New Zealand Herald*, 5 February 1931.
34 EPR, Mary Hunter memoir.
35 *New Zealand Herald*, 6 February 1931.
36 Ibid.
37 The hospital, Napier's second, opened on the old hilltop barracks site in 1880. Wright, *Hawke's Bay — The History of a Province*, pp. 72–73.
38 *New Zealand Herald*, 6 February 1931.
39 *Hawke's Bay — Before and After*, p. 76.
40 EPR, Jessie Atkinson memoir.
41 *New Zealand Herald*, 6 February 1931.
42 *New Zealand Herald*, 5 February 1931.
43 *Hawke's Bay — Before and After*, p. 67.
44 EPR, George Howard Brown memoir.
45 EPR, C.E. MacMillan, letter to his wife, 16 March 1931.
46 EPR, George Howard Brown memoir.
47 *Hawke's Bay — Before and After*, p. 85.
48 *New Zealand Herald*, 6 February 1931.
49 EPR, Major A.F.R. Irwin memoir.
50 Wright, *Hawke's Bay — The History of a Province*, pp. 124–125.
51 *Hawke's Bay — Before and After*, p. 87.
52 *New Zealand Herald*, 6 February 1931.
53 *Hawke's Bay — Before and After*, p. 89.
54 Ibid., p. 114.
55 *Taradale Town District Jubilee 1886–1936*, [n.d.], p. 31.
56 I.R. Robinson and H.L. Benjamin, 'Effects of earthquakes on electrical supply systems', Paper read at the Annual General Meeting of the New Zealand Society of Civil Engineers (Inc.), held at Christchurch, 21–25 February 1933, p. 8.
57 F.R. Callaghan, 'The Hawke's Bay Earthquake. General Description', in *New Zealand Journal of Science and Technology*, Vol. XV, No. 1, July 1933, p. 25.
58 Robinson & Benjamin, pp. 8–9.
59 'Hastings, Jewel of the Plains', pamphlet, 1938.
60 *New Zealand Herald*, 5 February 1931.
61 *Hawke's Bay — Before and After*, p. 73.
62 EPR, Vera Smith memoir.
63 *Hawke's Bay — Before and After*, p. 81.
64 Ibid.
65 Ibid., p. 78.
66 EPR, Vera Smith memoir.
67 *Hawke's Bay — Before and After*, p. 81.
68 *Hawke's Bay Tribune*, Earthquake Edition No. 1, 5 February 1931.
69 *New Zealand Herald*, 5 February 1931.
70 *Hawke's Bay — Before and After*, p. 72.
71 Ibid., p. 73.
72 *New Zealand Herald*, 4 February 1931.
73 For discussion see Wright, *Havelock North — The History of a Village*.
74 WTu, MS Papers 2418, Folder 1. W.J.C. & W.H. Ashcroft Collection, Bill Ashcroft to his sister, 15 February 1931.
75 EPR, Bernard Chambers diary extract; see also S.W. Grant, *In Other Days — A History of the Chambers Family of Te Mata, Havelock North*, CHB Printers, 1980, p. 137.
76 *Havelock North Primary School 125th Jubilee 1863–1988*, HNL A/335/1988, p. 42.
77 WTu, MS Papers 5814. Dorothy Campbell, letter to 'Aunt Frances', 5 March 1931.
78 Herbert Guthrie-Smith, *Tutira*, 3rd ed., Wm Blackwood & Sons, London, 1953, p. 47.
79 EPR, Darry McCarthy, transcript of radio talk, 3 February 1987.
80 *Hawke's Bay — Before and After*, p. 115.
81 Ibid., p. 123.
82 *New Zealand Herald*, 5 February 1931.
83 *Hawke's Bay — Before and After*, p. 115.
84 *New Zealand Herald*, 4 February 1931.
85 *New Zealand Herald*, 6 February 1931.
86 WTu, MS Papers 5283. Ernest St Clair Haydon papers.
87 H. Skinner, pers. comm.
88 *New Zealand Herald*, 4 February 1931.
89 Ibid.
90 Ibid.
91 M. Wynn, pers. comm.
92 WTu, MS Papers 6096. Hamilton Fellowes Baird, 'The Hawke's Bay Earthquake of 3rd February 1931' (narrative).
93 *New Zealand Herald*, 4 February 1931.
94 *New Zealand Herald*, 5 February 1931.

Chapter 2: 'Sand and water is not very sticky'

1 Charlotte Godley, *Letters from Early New Zealand*, Whitcombe & Tombs Ltd, Christchurch, 1951, p. 41.
2 Anna Rogers, *New Zealand Tragedies — Earthquakes*, Grantham House, Wellington, 1996, pp. 74–80.
3 WTu, MS Papers 1348. Frederick John Tiffen.
4 Godley, p. 41.
5 G.L. Downes, *Atlas of Isoseismal Maps of New Zealand Earthquakes*, Institute of Geological &

Nuclear Sciences Ltd, Lower Hutt, 1995, Plate 1. The location of the epicentre is open to discussion.
6 Downes, Plate 2.
7 Ibid; see also Rogers, pp. 81–87 for further details. In Wairoa the strength was approximately MM 5.5.
8 WTu, MS Papers 1348. Frederick John Tiffen.
9 R.I. Walcott, '. . . the gates of stress and strain . . .', in *Large Earthquakes in New Zealand*, Royal Society of New Zealand, Miscellaneous Series No. 5, Wellington, 1981, p. 13. Following Eiby (1971).
10 Wright, *Hawke's Bay — The History of a Province*, pp. 27–29; also P.J. Goldsmith, 'Aspects of the Life of William Colenso', MA thesis, University of Auckland, 1995.
11 The town was conceived in 1851 and designed in 1853. Wellington Provincial Superintendent Isaac Featherston blocked section sales until 1855. Matthew Wright, *Napier — City of Style*, Random House, Auckland, 1996, pp. 40–41.
12 Wright, *Havelock North — The History of a Village*, pp. 17–20.
13 *Hawke's Bay Herald*, 25 February 1863.
14 Ibid.
15 Wilson, p. 450.
16 NAMU, diary of H.W.P. Smith. Smith was an intermittent diary keeper and put the date at 21 March; context suggests 22 February.
17 Wilson, p. 450.
18 *Hawke's Bay Herald*, 21 March 1863.
19 *Hawke's Bay Herald*, 10 August 1904.
20 Downes, Plate 8.
21 Wright, *Havelock North — The History of a Village*, pp. 83–86, 178–180.
22 Principal literary exponents included Franz Kafka, Ezra Pound and Virginia Woolf. Modernist architects included Frank Lloyd Wright, Walther Gropius and Gerrit T. Rietveld. The concepts were replaced with 'post-modernism' after the Second World War.
23 The term 'art deco' originated with the 'decorative arts' school of Paris, popularised by a 1925 exhibition.
24 Most buildings are better defined with a date range; some years elapsed between concept, design and completion.
25 Wright, *Havelock North — The History of a Village*, pp. 83–88, 94–96, 173.
26 Peter Shaw, *Louis Hay: Architect*, Hawke's Bay Cultural Trust, Napier, 1999, pp. 38, 40. The fire station was designed in 1921 and completed in 1926.
27 Geoffrey Thornton, *Cast in Concrete*, Reed Publishing (NZ) Ltd, Auckland, 1996, p. 192.
28 WTu, MS 'Council of Fire and Accident Associations of New Zealand. Official records of Napier Earthquake, February 3rd–10th 1931', pp. 1–11.
29 Chapple, p. 5, notes that the depression bottomed out in 1933.
30 Laurie Barber, *New Zealand — A Short History*, Century Hutchinson, Auckland, 1989, p. 131. See also Wright, *Havelock North — The History of a Village*, p. 191.

Chapter 3: 'The glare ... lit up the sky'

1 EPR, Bernard Chambers diary extract.
2 Guthrie-Smith, p. 50.
3 F.C. Wright (1888–1976), pers. comm. Wright joined the Duke of Cornwall Light Infantry in 1906 and was seriously wounded at the first Battle of Ypres in 1914. He emigrated to Napier in 1922.
4 EPR, P.W. Barlow memoir.
5 EPR, R.S. Munro memoir.
6 WTu, MS 'Council of Fire and Accident Associations of New Zealand. Official records of Napier Earthquake, February 3rd–10th 1931'. Napier block plan summary.
7 *Hawke's Bay — Before and After*, p. 65.
8 WTu, MS 'Council of Fire and Accident Associations of New Zealand. Official records of Napier Earthquake, February 3rd–10th 1931'. Napier block plan summary.
9 *Hawke's Bay — Before and After*, p. 124.
10 Conly, pp. 46–47. In, e.g., *New Zealand Herald*, 5 February 1931, 'The Full Story of the Great Earthquake Disaster', *The Weekender*, 12 March 1931, etc., Edith Barry was listed as 'Mrs T. Barry', short for 'Mrs Tom Barry Senior'. Her full name was given on the official casualty register. *Hawke's Bay — Before and After*, p. 66.
11 *New Zealand Herald*, 5 February 1931.
12 Conly, p. 47; also *New Zealand Herald*, 5 February 1931.
13 *New Zealand Herald*, 5 February 1931.
14 Ibid.
15 Ibid.
16 *New Zealand Herald*, 6 February 1931.
17 Doctors practising in Napier at the time included Drs Barnett, Bernau, Birkenshaw, A. & H. Berry, Edgar, T.C. Moore, W.W. Moore, Oulton, Scoular, Waters, Waterworth and Will. *Hawke's Bay — Before and After*, p. 107.
18 EPR, J.S. Peel memoir.
19 *New Zealand Herald*, 6 February 1931.
20 EPR, Mary Eames memoir.
21 *New Zealand Herald*, 5 March 1931.
22 These stories applied to both Napier and Hastings, and proved persistent; see M.D.N. Campbell, *The Story of Napier*, Napier City Council, Napier, 1974, p. 132.
23 EPR, Mary Eames memoir.
24 EPR, Margaret [illegible] memoir.

25 EPR, George Brown memoir.
26 *New Zealand Herald*, 5 February 1931.
27 *Hawke's Bay —Before and After*, p. 85.
28 Marion Morris, 'Birthday present sent waves round the world', *Herald Tribune*, 3 February 1990.
29 WTu, MS Papers 3359. *Taranaki* (ship), radio logs.
30 Ibid.
31 Robinson & Benjamin, p. 6.
32 WTu, MS Papers 3359. *Taranaki* (ship), radio logs.
33 Ibid.
34 Ibid.
35 Ibid.
36 *New Zealand Herald*, 4 February 1931.
37 WTu MS 'Council of Fire and Accident Associations of New Zealand. Official records of Napier Earthquake, February 3rd–10th 1931'. Napier block plan summary.
38 EPR, Mary Hunter memoir.
39 Callaghan, p. 12.
40 WTu MS 'Council of Fire and Accident Associations of New Zealand. Official records of Napier Earthquake, February 3rd–10th 1931'. Napier block plan summary.
41 *Hawke's Bay — Before and After*, p. 71.
42 WTu, MS Papers 2107. Jean Anderson, letter to Elsie Young, 3 February 1931.
43 *Hawke's Bay — Before and After*, p. 71.
44 *Hawke's Bay — Before and After*, p. 124.
45 WTu, MS Papers 2418, Folder 1. W.J.C. & W.H. Ashcroft Collection.
46 *New Zealand Herald*; 9 February 1931. For pictures of the North Pond reclamation process see Wright, *Napier — City of Style*, pp. 68–69.
47 Callaghan, p. 15.
48 'How Earthquakes and Fires Wrought Destruction to Hastings, February 3rd 1931', pamphlet c1931, pp. 2–3.
49 EPR, Vera Smith memoir.
50 E.F. Scott, 'A Report on the relief organisation arising out of the earthquake in Hawke's Bay on February 3rd, 1931', Christchurch Public Utilities Committee, April 1931, p. 7.
51 Ibid., p. 9.
52 *New Zealand Herald*, 6 February 1931.
53 HNL, S.M.M. von Dadelszen, née Gardiner, earthquake memoir.
54 Scott, p. 4.
55 *New Zealand Herald*, 7 July 1931.
56 WTu, MS Papers 2418, Folder 1. W.J.C. & W.H. Ashcroft Collection, Bill Ashcroft to his sister, 15 February 1931.
57 Scott, p. 3.
58 *Hawke's Bay Tribune*, Earthquake Edition, No. 4, 7 February 1931.
59 Scott, p.2. This contemporary account varies from Mary Boyd, *City of the Plains*, Hastings City Council, Hastings, 1984, p. 260.
60 Ibid., pp. 2, 5.
61 Reports and minutes of meetings of council and other persons regarding earthquake matters, between the period of 3rd and 13th February 1931, in the temporary premises, Wesley Hall, Hastings: 3 February 1931. Hereafter noted as 'Minutes'.
62 Matthew Wright, *Kiwi Air Power*, Reed Publishing (NZ) Ltd, Auckland, 1998, p. 18. See also WTu, MS Papers 3359. *Taranaki* (ship).
63 *New Zealand Herald*, 5 February 1931.
64 WTu, MS Papers 2418, Folder 1. W.J.C. & W.H. Ashcroft Collection, Bill Ashcroft to his sister, 15 February 1931.
65 WTu, MS Papers 5814. Dorothy Campbell, letter to 'Aunt Frances', 5 March 1931.
66 WTu, Ms Papers 5283. Ernest St Clair Haydon Collection.
67 *Hawke's Bay — Before and After*, p. 123.
68 Guthrie-Smith, p. 48.
69 Ibid., pp. 49–51.
70 WTu, Ms Papers 5283. Ernest St Clair Haydon Collection.
71 *Hawke's Bay — Before and After*, p. 71. Napier has shingle beaches.
72 EPR, Mary Hunter memoir.
73 *New Zealand Herald*, 6 February 1931.
74 *New Zealand Herald*, 4 February 1931. The time in Wairoa was reported as 9.20 p.m.
75 *Hawke's Bay — Before and After*, p. 121.
76 EPR, Bernard Chambers diary extract.
77 WTu, MS Papers 2418, Folder 1. W.J.C. & W.H. Ashcroft Collection, Bill Ashcroft to his sister, 15 February 1931.
78 *Hawke's Bay — Before and After*, p. 87.
79 EPR, Mary Hunter memoir.

Chapter 4: 'Thank God for the Navy!'

1 Attributed to a Napier hill resident, 4 February 1931. *Hawke's Bay — Before and After*, p. 84. See also Minhinnick cartoon caption, *New Zealand Herald*, 9 February 1931.
2 *Hawke's Bay — Before and After*, p. 149.
3 *New Zealand Herald*, 4 February 1931.
4 *Hawke's Bay — Before and After*, p. 84.
5 *New Zealand Herald*, 4 February 1931.
6 *New Zealand Herald*, 6 February 1931.
7 *New Zealand Herald*, 4 March 1931.
8 *New Zealand Herald*, 5 March 1931.
9 Robinson & Benjamin, p. 9.
10 *New Zealand Herald*, 5 March 1931.
11 EPR, George Brown memoir.
12 *Hawke's Bay — Before and After*, p. 86.
13 Ibid., p. 150.

14 Ibid., p. 96.
15 *New Zealand Herald*, 6 February 1931.
16 WTu, MS Papers 5283. Ernest St Clair Haydon Collection.
17 *Hawke's Bay — Before and After*, p. 158.
18 Ibid., p. 105.
19 Ibid., p. 105.
20 Ibid., p. 91.
21 EPR, W.D. Corbett, City Engineer, memoir.
22 *Daily Telegraph* news bulletin, 4 March 1931.
23 *Hawke's Bay — Before and After*, p. 156.
24 *Hawke's Bay Tribune*, Earthquake Edition, No. 2, 5 February 1931.
25 *New Zealand Herald*, 4 February 1931.
26 WTu, MS Papers 3838. M.D. MacNab, 'Letter written February 1985 in response to a radio programme'.
27 WTu, MS Papers 3146. A.E.L. Bennett Collection.
28 *Hawke's Bay Tribune*, Earthquake Edition, No. 4, 7 February 1931.
29 WTu, MS Papers 3838. M.D. MacNab 'Letter written February 1985 in response to a radio programme'.
30 Ibid.
31 J. Wright, pers. comm.
32 *New Zealand Herald*, 5 March 1931.
33 *Hawke's Bay — Before and After*, p. 103.
34 Ibid., p. 94.
35 WTu, MS Papers 1346. A.E.L. Bennett Collection.
36 *Hawke's Bay — Before and After*, p. 117.
37 Scott, p. 15.
38 Ibid., p. 15.
39 Ibid., p. 18.
40 *Hawke's Bay — Before and After*, p. 97.
41 Callaghan, p. 31.
42 *Hawke's Bay — Before and After*, p. 121.
43 WTu, MS Papers 1346. A.E.L. Bennett Collection.
44 Callaghan, p. 31.
45 *Hawke's Bay —Before and After*, pp. 120–121.
46 Robinson & Benjamin, pp. 6, 9.
47 Callaghan, pp. 30–31.
48 *Hawke's Bay Tribune*, Earthquake Edition, No. 3, 6 February 1931.
49 *Hawke's Bay Tribune*, Earthquake Edition, No. 8, 11 February 1931.
50 Scott, p. 3.
51 *Daily Telegraph* news bulletin, 4 February 1931.
52 *Daily Telegraph* news bulletin, 5 February 1931.
53 *Hawke's Bay — Before and After*, p. 150.
54 *New Zealand Herald*, 6 March 1931.
55 WTu, MS Papers 5814. Dorothy Campbell letter to 'Aunt Frances', 5 March 1931.
56 *Hawke's Bay — Before and After*, p. 78.
57 *New Zealand Herald*, 5 February 1931.
58 WTu, MS Papers 5283. Ernest St Clair Haydon Collection.
59 Ibid.
60 *Daily Telegraph* news bulletin, 5 February 1931.
61 Callaghan, p. 33.
62 *New Zealand Herald*, 9 February 1931.
63 *Daily Telegraph* news bulletin, 7 February 1931.
64 Scott, p. 11.
65 *Daily Telegraph* news bulletin, 6 February 1931.
66 *Daily Telegraph* news bulletin, 13 February 1931.
67 WTu, MS Papers 5814. Dorothy Campbell, letter to 'Aunt Frances', 5 March 1931.
68 Thorns & Sedgwick, 1999, p. 132. H. Skinner, pers. comm., noted that no houses in Puketitiri were locked in the 1930s.
69 Some 37,214 offences were reported in 1930 in a national population of 1,506,800. This dropped to 33,168 offences in 1935. Rates climbed from 349,193 reported crimes in 1980 to 1,049,915 in 1994 in a national population of just over 3,325,900. Thorns & Sedgwick, pp. 131–132. Traffic offences contributed to the 1994 rate. Eric Hobsbawm, *Age of Extremes — The Short Twentieth Century, 1914–1991*, Abacus, 1994, especially pp. 49–51, 336–343, 565, argued that the twentieth century brought a general degradation of standards worldwide.
70 Scott, p. 5.
71 Ibid., p. 6.
72 *Hawke's Bay — Before and After*, p. 70.
73 Scott, p. 7.
74 WTu, MS Papers 2418, Folder 1. W.J.C. & W.H. Ashcroft Collection, Bill Ashcroft to his sister, 15 February 1931.
75 *Daily Telegraph* news bulletin, 6 February 1931.
76 WTu, MS Papers 5814. Dorothy Campbell, letter to 'Aunt Frances', 5 March 1931.
77 WTu, MS Papers 2418, Folder 1. W.J.C. & W.H. Ashcroft Collection, Bill Ashcroft to his sister, 15 February 1931.
78 *Daily Telegraph* news bulletin, 6 February 1931.
79 *Daily Telegraph* news bulletin, 7 February 1931.
80 Souvenir Issue of the *Hawke's Bay Tribune*, Emergency Earthquake Editions, HPL.
81 HNL, S.M.M. von Dadelszen, née Gardiner, earthquake memoir.
82 *Daily Telegraph* news bulletin, 6 February 1931.
83 Wright, *Napier — City of Style*, p. 82.
84 *New Zealand Herald*, 9 February 1931.
85 EPR, R.S. Munro memoir.
86 EPR, P.W. Barlow memoir.
87 EPR, Mary Hunter memoir.
88 This pavilion survived the earthquake with little damage and was demolished in the late 1990s.
89 *Hawke's Bay — Before and After*, p. 110.
90 WTu, MS Papers 1346. A.E.L. Bennett Collection.
91 *Hawke's Bay — Before and After*, p. 97.
92 *Daily Telegraph* news bulletin, 11 February 1931.

93 EPR, L.G. Grant memoir.
94 *New Zealand Herald*, 6 February 1931.
95 WTu, MS Papers 5814. Dorothy Campbell.
96 *Hawke's Bay — Before and After*, p. 77.
97 EPR, Mary Hunter memoir.
98 Chapple, p. 12. About a hundred vehicles were destroyed in the earthquake.
99 *Hawke's Bay — Before and After*, p. 118.
100 Ibid., p. 78.
101 EPR, Mary Hunter memoir.
102 EPR, George Brown memoir.
103 EPR, Rona Lawrence memoir.
104 *New Zealand Herald*, 9 February 1931.
105 *Hawke's Bay Tribune*, Earthquake Edition, No. 8, 11 February 1931.
106 WTu, MS Papers 5283. Ernest St Clair Haydon Collection.
107 Minutes, 5 February 1931.
108 *Hawke's Bay Tribune*, Earthquake Edition, No. 9, 12 February 1931.
109 Ibid.
110 HNL, S.M.M von Dadelszen, née Gardiner, earthquake memoir. The same story was recounted from a different perspective in the *Havelock North Village Press*, 27 July 2000.
111 WTu, MS Papers 5814. Dorothy Campbell letter to 'Aunt Frances', 5 March 1931.
112 *New Zealand Herald*, 9 February 1931; also *Hawke's Bay — Before and After*, p. 67.
113 *New Zealand Herald*, 5 February 1931.
114 EPR, unidentified note.
115 *Daily Telegraph* news bulletin, 11 February 1931.
116 *Daily Telegraph* news bulletin, 7 February 1931.
117 *New Zealand Herald*, 5 March 1931.
118 *New Zealand Herald*, 7 February 1931.
119 WTu, MS Papers 5283. Ernest St Clair Haydon Collection.
120 Callaghan, p. 37.
121 *New Zealand Herald*, 9 February 1931.
122 Callaghan, p. 37.
123 EPR, W. Olphert memoir.
124 WTu, MS Papers 1346. A.E.L. Bennett Collection.
125 Scott, p. 5.
126 *Hawke's Bay — Before and After*, p. 78.
127 *New Zealand Herald*, 6 February 1931.
128 *New Zealand Herald*, 5 February 1931.
129 WTu, MS Papers. Jean Anderson, letter to Elsie Young, 3 February 1931.
130 EPR, Mary Hunter memoir.
131 Grant Howard, *The Navy in New Zealand: An Illustrated History*, A.H. & A.W. Reed Ltd, Wellington, 1980, p. 139.
132 *Hawke's Bay — Before and After*, p. 77.
133 WTu, MS Papers 5814. Dorothy Campbell, letter to 'Aunt Frances', 5 March 1931.
134 HNL, S.M.M. von Dadelszen, née Gardiner, earthquake memoir.
135 WTu, MS Papers 5814. Dorothy Campbell, letter to 'Aunt Frances', 5 March 1931.
136 EPR, Bernard Chambers memoir.
137 Wright, *Havelock North — The History of a Village*, p. 180.
138 WTu, MS Papers 5283. Ernest St Clair Haydon Collection.
139 WTu, MS Papers 5814. Dorothy Campbell, letter to 'Aunt Frances', 5 March 1931.
140 WTu, MS Papers 1346. A.E.L. Bennett Collection.
141 *New Zealand Herald*, 9 February 1931.
142 *Hawke's Bay — Before and After*, p. 123.
143 *New Zealand Herald*, 14 February 1931.
144 Guthrie-Smith, p. 50.
145 Scott, p. 8.
146 *New Zealand Herald*, 14 February 1931.
147 Downes, Plate 15.
148 *New Zealand Herald*, 14 February 1931.
149 Ibid.

Chapter 5: '... an indescribable chaos of debris'

1 WTu, MS Papers 1346. A.E.L. Bennett Collection.
2 *New Zealand Herald*, 5 February 1931.
3 HPL Box 4 84/38, 'Earthquake Relief — Restoration Finance. Urgent Measures Required.' Reprinted from *New Zealand Financial Times*, February 1931.
4 WTu, MS Papers 5283. Ernest St Clair Haydon Collection.
5 EPR, C.E. MacMillan memoir.
6 WTu, MS Papers 5814. Dorothy Campbell, letter to 'Aunt Frances', 5 March 1931.
7 *Hawke's Bay — Before and After*, p. 153.
8 *Daily Telegraph* news bulletin, 11 February 1931.
9 WTu, MS Papers 2418, Folder 1. W.J.C. & W.H. Ashcroft Collection, Bill Ashcroft to his sister, 15 February 1931.
10 *Daily Telegraph* news bulletin, 9 February 1931.
11 *New Zealand Herald*, 9 February 1931.
12 Wilson, p. 375.
13 *Daily Telegraph* news bulletin, 7 February 1931.
14 *Daily Telegraph* news bulletin, 9 February 1931.
15 Scott, p. 8.
16 Ibid., p. 9.
17 *Hawke's Bay — Before and After*, p. 74.
18 *Daily Telegraph* news bulletin, 13 February 1931.
19 *Hawke's Bay — Before and After*, p. 156.
20 Robinson & Benjamin, p. 8.
21 Ibid., pp. 7–9.
22 *Hawke's Bay — Before and After*, p. 92.
23 See, e.g., Tony Simpson, *The Sugarbag Years*.
24 New Zealand Statutes, Geo. V, 1931, No. 6, 'The Hawke's Bay Earthquake Act, 1931'.
25 Callaghan, p. 35.
26 Wright, *Havelock North — The History of a Village*, p. 185.

27 Ibid.
28 Ibid., p. 186.
29 Callaghan, p. 37.
30 WTu MS 'Council of Fire and Accident Associations of New Zealand. Official records of Napier Earthquake, February 3rd–10th 1931', pp. 1–11.
31 *New Zealand Herald*, 4 February 1931.
32 *New Zealand Herald*, 5 February 1931.
33 WTu, MS Papers 3196, Sir Frederic Ponsonby; and MS Papers 3197, Sir Edward William Collection.
34 *Hawke's Bay — Before and After*, p. 86.
35 New Zealand Statutes, Geo. V, 1931, No. 29, 'The Hawke's Bay Earthquake Relief Fund Act, 1931'.
36 Campbell, p. 143.
37 Chapple, p. 11, following Dowrick et al. (1995).
38 Cited in Chapple, p. 11.
39 Chapple, p. 11, following Dowrick et al. (1995).
40 Dowrick et al. (1995), p. 135.
41 Chapple, p. 11, following Dowrick et al. (1995).
42 Author, personal observation.
43 HPL, Earthquake Relief Committee Correspondence.
44 Ibid.
45 WTu, MS Papers 5814. Dorothy Campbell, letter to 'Aunt Frances', 5 March 1931.
46 Cited in Wright, *Havelock North — The History of a Village*, p. 183.
47 EPR, Mary Hunter memoir.
48 Callaghan, p. 34.
49 Ibid., p. 37.
50 *Hawke's Bay — Before and After*, p. 94.
51 EPR, R.S. Munro memoir.
52 Callaghan, p. 35.
53 *Daily Telegraph*, 16 February 1931.
54 Claims noted and refuted by Chapple, p. 14.
55 *Hawke's Bay — Before and After*, p. 92.
56 Chapple, Figure 1, p. 14.
57 Callaghan, p. 37.
58 Chapple, p. 14, which differs from Callaghan's figures.
59 Callaghan, p. 37.
60 Ibid., p. 35.
61 Wright, *Napier — City of Style*, p. 100.
62 Dowrick (1998), pp. 157, 163.
63 These figures noted by Dowrick (1998) supersede Boyd, p. 267.
64 WTu, MS Papers 6096. Hamilton Fellowes Baird, 'The Hawke's Bay Earthquake of 3rd February 1931'.
65 WTu, MS Papers 2418, Folder 1. W.J.C. & W.H. Ashcroft Collection, Bill Ashcroft to his sister, 15 February 1931.
66 Callaghan, p. 37, cites these permits as being worth £339,538 ($24,446,736), but his figures have been revised by Chapple, p. 14.
67 Noel C. Harding, 'Hastings (NZ) from Town Board to City 1884–1962', typescript, 1962, pp. 56–57.
68 Harding, p. 58.
69 *Hawke's Bay Tribune*, Earthquake Edition, No. 7, 10 February 1931.
70 WTu, MS Papers 6096. Hamilton Fellowes Baird, 'The Hawke's Bay Earthquake of 3rd February 1931'.
71 *Hawke's Bay Tribune*, Earthquake Edition, No. 8, 11 February 1931.
72 Hobsbawm, *Age of Extremes*, pp. 84–107, argues that the depression almost destroyed the capitalist system.
73 Callaghan, pp. 34–37.
74 Chapple, p. 48. Hobsbawm, *Age of Extremes*, pp. 86–87, 102, makes useful general commentary on the 1930s historical context of neo-classical theory, notably: 'Those of us who lived through the years of the Great Slump still find it almost impossible to understand how the orthodoxies of the pure free market, then so obviously discredited, once again came to preside over a global period of depression in the 1980s and 1990s which, once again, they were equally unable to understand or deal with.'
75 Chapple, pp. 37–38, citing *New Zealand Yearbooks*.
76 Boyd, p. 270, cites 600; Chapple, p. 44, suggests 300.
77 Figures from Chapple, p. 44; my interpretation differs from Chapple.
78 Boyd, p. 274.
79 See Chapple, p. 6; also Thorns & Sedgwick, p. 64.
80 Chapple, p. 8.
81 Chapple, p. 25. Following Callaghan (1933), Chapple did not take business losses in Havelock North into account. His top figure is nevertheless likely to be in the right 'ballpark'.
82 Dowrick (1998), pp. 139–140.

Chapter 6: An unsteady legacy

1 J. Henderson, 'Geological aspects of the Hawke's Bay earthquake', in *New Zealand Journal of Science and Technology*, Vol. XV, No. 1, July 1933.
2 *New Zealand Herald*, 6 February 1931.
3 See Hobsbawm, *Age of Extremes*, pp. 549–550, for discussion of historical context.
4 For a summary of the mechanisms see Aitken, pp. 14–20.
5 The South Island system involves different plate interactions and has created an earthquake zone extending into the North Island system.
6 A.J. Bevin, 'Geodetic Surveys for Earth Deformation', in *Large Earthquakes in New Zealand*, Royal Society of New Zealand, Miscellaneous Series No. 5, Wellington, 1981, p. 87.

7 Malcolm McKinnon (ed.), *New Zealand Historical Atlas*, David Bateman, Auckland, 1997, plates 4–6, provides a summary.
8 Other geological processes also influence landforms.
9 Walcott, '. . . the gates of stress and strain . . .', in *Large Earthquakes in New Zealand*, p. 16.
10 Aitken, p. 81.
11 Guthrie-Smith, p. 46.
12 Downes, p. 4, also Plate 16.
13 Dowrick et al. (1995), p. 141, after Haines & Darby (1987).
14 Dowrick et al. (1995), p. 135.
15 WTu, MS Papers 6096. Hamilton Fellowes Baird, 'The Hawke's Bay Earthquake of 3rd February 1931'.
16 D.J. Dowrick and E.G.C. Smith, 'Surface wave magnitudes of some New Zealand earthquakes 1901–1988', *Bulletin of the New Zealand National Society for Earthquake Engineering*, Vol. 23, No. 3, September 1990, p. 209.
17 Aitken, pp. 31–38.
18 Dowrick & Smith.
19 See, e.g., Dowrick (1998).
20 Downes, pp. 12–17. Also Warwick D. Smith, 'The Vast Event — How Vast and How Often? A statistical perspective of earthquake occurrence', in *Large Earthquakes in New Zealand*, Royal Society of New Zealand, Miscellaneous Series No. 5, Wellington, 1981, pp. 22–23.
21 Dowrick (1998), esp. pp. 155–158. Chimneys were observed falling in Wellington in Tinakori Road, near the Wellington fault line, indicating that here the Hawke's Bay quake had a felt intensity of up to VI on the Modified Mercalli Scale. M. Wynn, pers. comm.
22 *Hawke's Bay — Before and After*, pp. 95–96.
23 Ibid., p. 134.
24 Ibid., p. 135.
25 Downes, Plate 23.
26 Ibid., Plate 25.
27 Author, personal observation.
28 Rogers, p. 149.
29 Ibid., pp. 160–162.
30 Ibid., p. 151. One individual died from heart failure.
31 *Hawke's Bay — Before and After*, p. 69.
32 Ibid., p. 100.
33 Ibid., p. 74.
34 *New Zealand Herald*, 5 February 1931.

Bibliography

PRIMARY SOURCES

Alexander Turnbull Library, National Library of New Zealand (WTu)

MANUSCRIPTS AND PAPERS

MS Papers 1346 A.E.L. Bennett Collection.
MS Papers 2107 Jean Anderson, letter to Elsie Young, 3 February 1931.
MS Papers 2418 W.J.C. and W.H. Ashcroft Collection.
MS Papers 2883 G.F. Marshall White, 'Notes on History of Hawke's Bay'.
MS Papers 3196 Sir Frederic Ponsonby, 'King's Quake Relief'.
MS Papers 3359 SS *Taranaki* (ship).
MS Papers 3838 M.D. MacNab Collection.
MS Papers 5283 Ernest St Clair Haydon Collection.
MS Papers 5814 Dorothy Campbell Collection.
MS Papers 6096 Hamilton Fellowes Baird Collection, 'The Hawke's Bay Earthquake of 3rd February 1931'.

PHOTOGRAPHIC ARCHIVE

'Council of Fire and Accident Associations of New Zealand. Official records of Napier Earthquake, February 3rd–10th 1931'.

Hastings Public Library (HPL)

63/100 'Reconstruction — Ten Months after New Zealand's Greatest Disaster'.
63/113 'Hastings: Health, Wealth and Prosperity, "The Jewel of the Plains"', c. 1939.
84/38 'Earthquake Relief — Restoration Finance. Urgent Measures Required'.
84/4 'How New Zealand's Greatest Disaster Shattered Hastings, Earthquakes and Fires, February 3rd 1931'.
98/45 Reports and minutes of meetings of council and other persons regarding earthquake matters between the period of 3rd and 13th February 1931, in the temporary premises, Wesley Hall, Hastings.
98/9 Souvenir Issue, *Hawke's Bay Tribune*, Emergency Earthquake Editions.
Earthquake relief committee correspondence.

Havelock North Public Library (HNL)

Hunt, Zelma, 'The History of Taradale', typescript.
von Dadelszen, S.M.M. (née Gardiner) memoir.
Havelock North Primary School 125th Jubilee 1863–1988.

Hawke's Bay Museum Library (NAMU)

Earthquake Personal Reminiscences Box 1 (EPR): Jessie Atkinson, P.W. Barlow, George Howard Brown, Joyce Bourgeois, Bernard Chambers (diary extracts), W.C. Corbett, Mary Eames, L.G. Grant, Mary Hunter, A.F.R. Irwin, Mitchell Des Landes, Rona Lawrence, Darry McCarthy, C.E. MacMillan letter to his wife, C.A. Mountfort, R.S. Munro, Jock Stevenson, W. Olphert, J.S. Peel, Vera Smith, Margaret [illegible].
Diary of H.W.P. Smith.

Hastings District Council Archive (HDC)

Harding, Noel C., 'Hastings (NZ) from Town Board to City 1884–1962', typescript, 1962.

Newspapers

Daily Telegraph
Hawke's Bay Herald Tribune
Hawke's Bay Herald
New Zealand Herald
Press
Evening Post
Wellington Independent

SECONDARY SOURCES

—*The Hawke's Bay Earthquake*, Daily Telegraph, Napier, 1989 (reprint 1998).
—*Hawke's Bay — Before and After*, Daily Telegraph, Napier, 1931 (reprint 1981).
—*New Zealand's Greatest Disaster — Shattered Hastings, Earthquakes and Fires, February 3rd 1931* (n.d.).
—*Taradale Town District Jubilee 1886–1936 Souvenir Booklet*, Taradale, 1936.
Adams, C.E., Barnett, M.A.F., and Hayes, R.C., 'Seismological report on the Hawke's Bay Earthquake', in *New Zealand Journal of Science and Technology*, Vol. XV, No. 1, July 1933.
Aitken, Jefley J., *Rocked and Ruptured: Geological Faults in New Zealand*, Reed Publishing (NZ) Ltd, Auckland, 1999.
Anderson, Len, *Coaches North*, A.H. & A.W. Reed, Wellington, 1967.
Barber, Laurie, *New Zealand — A Short History*, Century Hutchinson, Auckland, 1989.
Bevin, A.J., 'Geodetic Surveys for Earth Deformation', in *Large Earthquakes in New Zealand*, Royal Society

of New Zealand, Miscellaneous Series No. 5, Wellington, 1981.

Boyd, M.B., *City of the Plains, A History of Hastings*, VUW Press/Hastings City Council, Wellington, 1984.

Callaghan, F.R., 'The Hawke's Bay Earthquake. General Description', in *New Zealand Journal of Science and Technology*, Vol. XV, No. 1, July 1933.

Campbell, M.D.N., *The Story of Napier*, Napier City Council, Napier, 1974.

Chapple, Simon, *The Economic Effects of the 1931 Hawke's Bay Earthquake*, New Zealand Institute of Economic Research (Inc.), Working Paper 97/7, Wellington, August 1997.

Conly, Geoff, *The Shock of '31 — The Hawke's Bay Earthquake*, A.H. & A.W. Reed, Wellington, 1980.

Conly, Geoff, *A Case History — The Hawke's Bay Hospital Board 1876–1989*, Hawke's Bay Area Health Board, Napier, 1992.

Cooney, R.C. and Fowkes, A.H.R., 'New Zealand Houses in Earthquakes — What Will Happen?', in *Large Earthquakes in New Zealand*, Royal Society of New Zealand, Miscellaneous Series No. 5, Wellington, 1981.

Downes, G.L., *Atlas of Isoseismal Maps of New Zealand Earthquakes*, Institute of Geological & Nuclear Sciences Ltd, Monograph 11, Lower Hutt, 1995.

Dowrick, David J, 'Damage and intensities in the magnitude 7.8 1931 Hawke's Bay, New Zealand earthquake', *Bulletin of the New Zealand National Society for Earthquake Engineering*, Vol. 31, No. 3, September 1998.

Dowrick, D.J., Rhoades, D.A., Babor, J., and Beetham, R.D., 'Damage ratios and microzoning effects in Napier in the magnitude 7.8 Hawke's Bay, New Zealand earthquake of 1931', *Bulletin of the New Zealand National Society for Earthquake Engineering*, Vol. 28, No. 2, June 1995.

Dowrick, D.J. and Smith, E.G.C., 'Surface wave magnitudes of some New Zealand earthquakes 1901–1988', *Bulletin of the New Zealand National Society for Earthquake Engineering*, Vol. 23, No. 3, September 1990.

Galloway, Ann, *Art Deco Napier — A Design Guide*, Napier City Council, Napier, 1992.

Godley, Charlotte, *Letters from Early New Zealand*, Whitcombe & Tombs Ltd, Christchurch, 1951.

Grant, S.W., *Havelock North, 1860–1952*, HB Newspapers, Hastings, 1978.

Grant, S.W., *In Other Days — A History of the Chambers Family of Te Mata, Havelock North*, CHB Printers, Waipukurau, 1980.

Guthrie-Smith, H., *Tutira* (3rd ed.), Wm Blackwood & Sons, London, 1953.

Henderson, J., 'Geological aspects of the Hawke's Bay earthquake', *New Zealand Journal of Science and Technology*, Vol. XV, No. 1, July 1933.

Hill, Elizabeth, *Between the Rivers*, CHB Printers & Publishers, Waipukurau, 1990.

Hobsbawm, Eric, *Age of Extremes — The Short Twentieth Century 1914–1991*, Abacus, London, 1995.

Hobsbawm, Eric, *On History*, Abacus, London, 1997.

Ives, Heather, *The Art Deco Architecture of Napier*, Ministry of Works and Development, Wellington, 1982.

Lloyd-Pritchard, M.F., *An Economic History of New Zealand to 1939*, Collins, Auckland, 1970.

Lovell-Smith, H.J., *Hastings & Hawke's Bay: New Zealand in Picture and Story*, Private Publication, c. 1935.

Main, Genifer (ed.), *Iona College: A Chronicle of 75 Years*, Taradale, 1989.

McDougall, R.J., *New Zealand Naval Vessels*, G.P. Books, Wellington, 1989.

McKinnon, Malcolm (ed.), *New Zealand Historical Atlas*, David Bateman, Auckland, 1997.

Morris, Marion, 'Birthday present sent waves round the world', *Herald Tribune*, 3 February 1990.

Roberts, J.L., 'The Reduction of Earthquake Hazards', in *Large Earthquakes in New Zealand*, Royal Society of New Zealand, Miscellaneous Series No. 5, Wellington, 1981.

Robinson, I.R. and Benjamin, H.L., 'Effects of earthquakes on electrical supply systems', paper read at the Annual General Meeting of the New Zealand Society of Civil Engineers (Inc.), 21–25 February 1933.

Rogers, Anna, *New Zealand Tragedies — Earthquakes*, Grantham House, Wellington, 1996.

Scott, Edwin F., 'A Report on the Relief Organisation arising out of the earthquake in Hawke's Bay on February 3rd, 1931', Christchurch Public Utilities Committee, April 1931.

Shaw, Peter, *Louis Hay: Architect*, Hawke's Bay Cultural Trust, Napier, 1991.

Simpson, Tony, *The Sugarbag Years*, Hodder & Stoughton, Auckland, 1974, new edition 1984.

Suggate, R.P. (ed.), *The Geology of New Zealand*, Government Printer, Wellington, 1978.

Thorns, David and Sedgwick, Charles, *Understanding Aotearoa/New Zealand: Historical Statistics*, Dunmore Press, Palmerston North, 1997.

Walcott, R.I., '... the gates of stress and strain ...' in *Large Earthquakes in New Zealand*, Royal Society of New Zealand, Miscellaneous Series No. 5, Wellington, 1981.

Wilson, J.G. (ed.), *History of Hawke's Bay*, A.H. & A.W. Reed, Wellington, 1939.

Wright, Matthew, *A History of the Eastern Kaweka Ranges*, NZFS, Napier, 1984.

—*The History of Farming at Kuripapango*, NZFS, Napier, 1984.

—*The Early Sawmilling Industry in Hawke's Bay*, Vol. 1, NZFS, Napier, 1985.

—*Hawke's Bay —The History of a Province*, Dunmore Press, Palmerston North, 1994.
—*Havelock North — The History of a Village*, Hastings District Council, Hastings, 1996.
—*Napier — City of Style*, Random House, Auckland, 1996.
—*Kiwi Air Power — The History of the RNZAF*, Reed Publishing (NZ) Ltd, Auckland, 1998.
—'City of style rose quickly from the rubble', *Daily Telegraph*, 2 February 1991.
—'After the quake', *Dominion Sunday Times*, 3 February 1991.
—'Napier toasts its heritage', *Evening Post*, 3 February 1992.
—'From shake to slump to merger', *Daily Telegraph*, 21 March 1992.
—'Celebration inspired by Napier's Art Deco heritage', *Dominion Sunday Times*, 30 January 1994.
—'Pre-quake Napier had its own style', *Impact*, February 1994.
—'New styles entrenched well before quake', *Daily Telegraph*, 15 February 1994.
—'Recreating the golden age', *Evening Post*, 13 February 1995.
—'Art Deco heritage feted by city', *Press*, 15 February 1995.
—'Napier fetes its flirtation with California', *Otago Daily Times*, 27 January 1996.
—'Early quake "as fierce a disaster as '31"', *Daily Telegraph*, 3 February 1996.

Index

About the Author

Matthew Wright is a professional historian and the author of more than 350 feature articles and numerous books. He has degrees in history and anthropology, and wrote his MA thesis on aspects of New Zealand's naval defence policy. He has been writing for many years on subjects ranging from defence policies to colonial social history, engineering history, land claim issues and forest industries. He won an award in 1996 for his interpretation of New Zealand provincial history, and pursues an active interest in New Zealand's military history.